Deep Learning
for EEG-Based
Brain–Computer
Interfaces

Representations, Algorithms and Applications

Deep Learning for EEG-Based Brain–Computer Interfaces

Representations, Algorithms and Applications

Xiang Zhang
Harvard University, USA

Lina Yao
University of New South Wales, Australia

World Scientific

NEW JERSEY · LONDON · SINGAPORE · BEIJING · SHANGHAI · HONG KONG · TAIPEI · CHENNAI · TOKYO

Published by

World Scientific Publishing Europe Ltd.

57 Shelton Street, Covent Garden, London WC2H 9HE

Head office: 5 Toh Tuck Link, Singapore 596224

USA office: 27 Warren Street, Suite 401-402, Hackensack, NJ 07601

Library of Congress Cataloging-in-Publication Data

Names: Zhang, Xiang, author. | Yao, Lina, author.

Title: Deep learning for EEG-based brain–computer interfaces :
    representations, algorithms and applications / Xiang Zhang, Harvard University, USA,
    Lina Yao, University of New South Wales, Australia.

Description: New Jersey : World Scientific, [2022] | Includes bibliographical references and index.

Identifiers: LCCN 2021024493 | ISBN 9781786349583 (hardcover) |
    ISBN 9781786349590 (ebook) | ISBN 9781786349606 (ebook other)

Subjects: LCSH: Brain-computer interfaces. | Machine learning.

Classification: LCC QP360.7 .Z43 2022 | DDC 612.8/20285--dc23

LC record available at https://lccn.loc.gov/2021024493

British Library Cataloguing-in-Publication Data

A catalogue record for this book is available from the British Library.

For any available supplementary material, please visit
https://www.worldscientific.com/worldscibooks/10.1142/Q0282#t=suppl

Typeset by Stallion Press
Email: enquiries@stallionpress.com

Preface

Research on brain signals is fascinating; it's like controlling the world with your mind. One big challenge obstructing your superpower is the ineffectiveness and inaccuracy of brain signal decoding. Fortunately, in recent years, powerful deep learning algorithms have achieved great success in a broad field of research through excellent high-level representation learning. In this book, we present how deep learning techniques enhance brain–computer interface (BCI) research.

To begin with, the book presents a new taxonomy of BCI signal paradigms according to the acquisition methods and systemically introduces the fundamental knowledge of deep learning models. Moreover, several guidelines for the investigation and design of BCI systems are provided from the aspects of signal category, deep learning models, and applications.

Special attention has been given to the state-of-the-art studies on deep learning for EEG-based BCI research in terms of algorithms. Specifically, this book introduces a number of advanced deep learning algorithms and frameworks aimed at several major issues in BCI including robust brain signal representation learning, cross-scenario classification, and semi-supervised classification.

Furthermore, several novel prototypes of deep learning-based BCI systems are proposed which shed light on real-world applications such as authentication, visual reconstruction, language interpretation, and neurological disorder diagnosis. Such applications can dramatically benefit both healthy individuals and those with disabilities in real life.

Xiang Zhang, Lina Yao
June 2020

Contents

Preface   v

Part 1: Background   **1**

1. Introduction   3

    1.1   Background on the Brain–Computer Interface   3
    1.2   Why Deep Learning?   5
    1.3   Why This Book?   6

2. Brain Signal Acquisition   9

    2.1   Invasive Approaches   9
          2.1.1   Intracortical Approaches   12
          2.1.2   Electrocorticography   13
    2.2   Noninvasive Approaches   14
          2.2.1   Electroencephalography   14
          2.2.2   Functional Near-infrared Spectroscopy   17
          2.2.3   Functional Magnetic Resonance Imaging   18
          2.2.4   Electrooculography   19
          2.2.5   Magnetoencephalography   20
    2.3   EEG Paradigms   21
          2.3.1   Spontaneous EEG   22
          2.3.2   Evoked Potential 22

3. Deep Learning Foundations 27

   3.1 Discriminative Deep Learning Models 29
      3.1.1 Multilayer Perceptron 29
      3.1.2 Recurrent Neural Networks 31
      3.1.3 Convolutional Neural Networks 34
   3.2 Representative Deep Learning Models 35
      3.2.1 Autoencoder 36
      3.2.2 Restricted Boltzmann Machine 38
      3.2.3 Deep Belief Networks 39
   3.3 Generative Deep Learning Models 40
      3.3.1 Variational Autoencoder 40
      3.3.2 Generative Adversarial Networks 42
   3.4 Hybrid Models 43

Part 2: Deep Learning-Based BCI and
        Its Applications **45**

4. Deep Learning-Based BCI 47

   4.1 Intracortical and ECoG 47
   4.2 EEG Potentials 48
      4.2.1 Spontaneous EEG Potentials 48
      4.2.2 Evoked Potentials 58
   4.3 fNIRS 61
   4.4 fMRI 62
   4.5 EOG 63
   4.6 MEG 64
   4.7 Discussion 64
      4.7.1 Discussions on Brain Signals 71
      4.7.2 Discussions on Deep Learning Models 73

5. Deep Learning-Based BCI Applications 77

   5.1 Health Care 77
   5.2 Smart Environments 84
   5.3 Communication 85
   5.4 Security 85
   5.5 Affective Computing 86
   5.6 Driver Fatigue Detection 86

5.7   Mental Load Measurement   87
5.8   Other Applications   88
5.9   Benchmark Data Sets   88
5.10 Discussions   91

**Part 3: Recent Advances on Deep Learning
          for EEG-Based BCI   95**

6. Robust Brain Signal Representation Learning   97

    6.1   Overview   97
    6.2   Subject-Dependent 100
          6.2.1   Temporal Representation Learning 100
          6.2.2   Spatial Representation Learning 103
          6.2.3   Graphical Representation Learning 105
          6.2.4   Spatiotemporal Representation Learning 109
          6.2.5   Discussion 112
    6.3   Cross-Subject 113
          6.3.1   Overview 113
          6.3.2   EEG Characteristic Analysis 113
          6.3.3   Representation Learning Framework 115
    6.4   Subject-Independent 120
          6.4.1   Transfer Learning 121
          6.4.2   Intersubject Transfer Learning 121

7. Cross-Scenario Classification   123

    7.1   Overview 123
    7.2   Attention-Based Classification Across Signal Sources . . . 124
          7.2.1   Overview 124
          7.2.2   Reinforced Selective Attention Model 125
          7.2.3   Discussion 134
    7.3   Attention-Based Classification Across Applications 135
          7.3.1   Overview 135
          7.3.2   Reinforced Attentive CNN 136
          7.3.3   Evaluation Across Applications 138
          7.3.4   Discussion 147
    7.4   Transfer Learning Methods 147

8. Semi-Supervised Classification   149

    8.1 Generative Methods 149
         8.1.1 Overview 149
         8.1.2 Adversarial Variational Embedding Algorithm . . 152
         8.1.3 Evaluation 157
         8.1.4 Discussion 161
    8.2 Wrapper Methods 161
         8.2.1 Self-Training 162
         8.2.2 Co-Training 163
         8.2.3 Boosting 164
    8.3 Unsupervised Representations Learning 164

Part 4: Typical Deep Learning for EEG-Based BCI Applications   167

9. Authentication   169

    9.1 EEG-Based Person Identification 169
         9.1.1 Challenges 169
         9.1.2 EEG Pattern Analysis 173
         9.1.3 Methodology 175
         9.1.4 Discussions 181
    9.2 Person Authentication 182
         9.2.1 Motivations 183
         9.2.2 Methodology 184
         9.2.3 Data Acquisition 189

10. Visual Reconstruction   191

    10.1 Brain2Object: Printing Your Mind 191
          10.1.1 Brain2Object System 192
          10.1.2 Data Acquisition 198
          10.1.3 Online System 199
          10.1.4 Discussions 201
    10.2 Geometrical Shape Reconstruction 202
          10.2.1 EEG Signal Acquisition 203
          10.2.2 Methodology 204
          10.2.3 Evaluations 208
          10.2.4 Discussions 210

11. Language Interpretation   211

     11.1 Methodology   212
           11.1.1 Overview   212
           11.1.2 Deep Feature Learning 212
           11.1.3 Feature Adaptation 215
     11.2 Brain-Controlled Typing System   216
     11.3 Discussion 219

12. Intent Recognition in Assisted Living   221

     12.1 System Overview   221
     12.2 Orthogonal Array Tuning Method   222
           12.2.1 Overview 222
           12.2.2 OATM Workflow   223
     12.3 Deployment 225
           12.3.1 Mind-Controlled Mobile Robot 225
           12.3.2 Mind-Controlled Appliances   226

13. Patient-Independent Neurological Disorder Detection   227

     13.1 Introduction 227
     13.2 Methodology 229
           13.2.1 Overview 229
           13.2.2 EEG Decomposition   231
           13.2.3 Attention-Based Seizure Diagnosis   233
           13.2.4 Patient Detection 234
           13.2.5 Training Details 235
     13.3 Discussions 235

14. Future Directions and Conclusion   237

     14.1 Future Directions   237
           14.1.1 General Framework 237
           14.1.2 Subject-Independent Classification 238
           14.1.3 Semi-Supervised and Unsupervised
                   Classification 238
           14.1.4 Hardware Portability   239
     14.2 Conclusion 240

Bibliography   241

Index   273

PART 1

Background

Chapter 1

Introduction

1.1 Background on the Brain–Computer Interface

BCI[1] systems translate human brain patterns into messages or commands for transmission to the outer world (Lotte *et al.*, 2018). BCI underpins many novel applications that are important in daily life, especially to individuals with psychological/physical or disabilities. On one hand, BCI can assist the disabled, the elderly, and people with limited motion ability (e.g., people with muscle diseases) in controlling wheelchairs, home appliances, and robots. For instance, a patients' brain signals can control household appliances through a BCI system. On the other hand, ordinary individuals can enjoy enhanced entertainment and security when brain waves–based techniques are involved (Zhang *et al.*, 2018g). Generally, a BCI system enables bidirectional communication between the human brain and the computer. However, as introduced in Abdulkader *et al.* (2015) and Roy *et al.* (2019), systems based on brain signal analysis (such as mental disease diagnosis, emotional computation, sleeping state scoring, etc.) can also be considered as more generalized forms of BCI. This book investigates the generalized BCI systems to provide a more comprehensive understanding of and scope for BCI systems.

Figure 1.1 shows the general paradigm of a BCI system, comprised of five key components: brain signal collection, signal preprocessing, feature

---

[1]Apart from BCI, there are a number of similar terms to define system in which machines are directly controlled by human brain signals, like Brain Machine Interface (BMI), Brain Interface (BI), Direct Brain Interface (DBI), Adaptive Brain Interface (ABI), and so on.

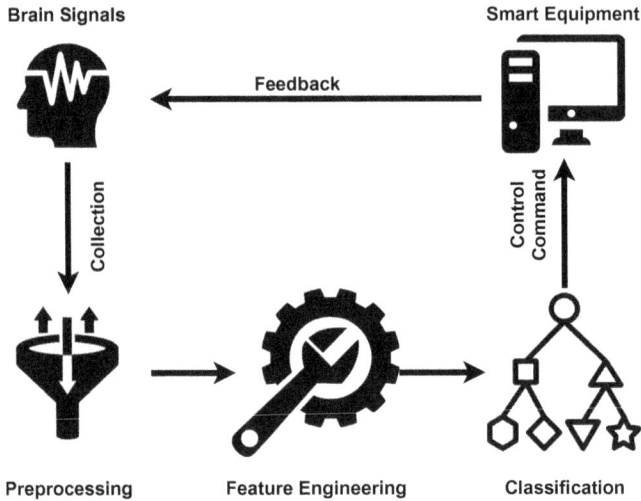

Fig. 1.1: General workflow of brain–computer interface system.

engineering, and smart equipment. First, human brain signals are collected from humans and sent to the preprocessing component for denoising and enhancement. Next, the discriminating features are extracted from the processed signals and sent to the classifier for further analysis. The classifier converts the decoded brain signals into digital commands to control the smart equipment and respond to the user.

The collection method, which is the first step of the procedure, is determined by the type of signal. For example, collection of EEG signals, which measure the voltage fluctuation resulting from ionic current within the neurons of the brain, requires placing a series of electrodes on to the scalp to record the electrical brain activity. Since the ionic current generated within the brain is measured at the scalp, intervening matter (e.g., the skull) greatly decrease the signal quality. Therefore, brain signals are usually preprocessed before feature engineering to increase the Signal-to-Noise Ratio (SNR). The preprocessing component comprises multiple steps such as signal cleaning (e.g., smoothing the noisy signals or resolving the inconsistencies), signal normalization (e.g., normalizing each channel of the signals along the time axis), signal enhancement (e.g., removing direct current), and signal reduction (presenting a reduced representation of the signal).

Feature engineering refers to the process of extracting discriminating features from the input signals using domain knowledge. Traditional features are extracted from the time domain (e.g., variance, mean value, kurtosis), frequency domain (e.g., fast Fourier transform), and time–frequency domains (e.g., discrete wavelet transform), and analyzed to enrich distinguishable information regarding user intention. Feature engineering is highly dependent on the domain knowledge. For example, biomedical knowledge is required to extract the features from brain signals which indicate epileptic seizures. Additionally, manual feature extraction is also time-consuming and may be inaccurate. Recently, automatic extraction of distinguishable features through deep learning has become a more attractive option.

The classification component refers to the machine learning algorithms that classify the extracted features into logical control signals recognizable by external devices. Recent studies have shown deep learning algorithms to be more powerful than traditional classifiers.

1.2 Why Deep Learning?

Although BCI systems have made tremendous progress (Abdulkader *et al.*, 2015; Bashashati *et al.*, 2007) over the past decades, the research still faces significant challenges.

First, brain signals are easily corrupted by various biological (e.g., eye blinks, muscle artifacts, fatigue, and concentration level) and environmental artifacts (e.g., noises) (Abdulkader *et al.*, 2015). As a result, the collected brain signals face the challenge of low SNR (Samek *et al.*, 2012). The low SNR cannot be easily addressed by traditional methods due to the time they require and the accompanying risk of information loss (Zhang *et al.*, 2018h). It is therefore crucial to be able to derive informative representations from corrupted brain signals.

Second, manual feature engineering is highly dependent on human expertise in the relevant domain. For example, basic biological knowledge is required to meaningfully investigate the sleeping state through EEG signals. General life experience may prove helpful in certain aspects but fall short overall. Manual analysis also faces time constraint issues due to the relatively low availability of expert investigators. Consequently, a novel method that can automatically extract the most representative features from the input data is highly desirable.

Third, most existing machine learning research focuses on static data and therefore, cannot accurately classify rapidly changing brain signals. For instance, the state-of-the-art classification accuracy for multiclass motor imagery EEG is generally below 80% (Lotte *et al.*, 2007). Dealing with dynamic data streams in BCI systems requires novel learning methods.

Furthermore, in practice, the annotated brain signal samples are not easily obtained, because such labeling requires a large number of certified labelers with professional knowledge. In some cases, therefore, the designer of the BCI systems will only have access to a small set of well-labeled brain signals, while the rest will be unlabeled samples. Meanwhile, improving the performance of BCI systems by including the information contained within the unlabeled data is crucial.

This book considers solutions to the aforementioned challenges through the use of deep learning. A subfield of machine learning, deep learning learns informative representations from the input data by building hierarchical networks (Goodfellow *et al.*, 2016). Deep learning can learn high-level and complicated representations by combining a number of simpler representations while each simple representation extracts partial information relating to the complex representation. In recent years, deep learning algorithms have been applied extensively in brain signal applications with great success (Cecotti *et al.*, 2014; Mahmud *et al.*, 2018). Compared to traditional machine learning methods, deep learning has several advantages. First, it works directly on raw brain signals, thus avoiding the time-consuming preprocessing and feature engineering in conventional BCI systems. Therefore, it eliminates much manual effort and alleviates the dependence on professional knowledge. Moreover, empowered by back-propagation, deep learning models are able to learn the "correct" message from brain signals. In other words, deep learning can learn high-quality representations even though the input is corrupted and noisy. Additionally, deep neural networks can capture both representative high-level features and latent dependencies through deep structures. More details of deep learning models will be introduced in Chapter 3.

1.3 Why This Book?

This book is believed useful for researchers and students in a wide range of fields: computer science, artificial intelligence, neuroscience, and others. We inspect and summarize most of the existing works that adopting deep learning algorithms to solve BCI problems. Then, we extract and present the crucial rules and typical algorithms/framework to effectively learn high-level representation from brain signals. In this book, we present both the basic knowledge of BCI systems and deep learning algorithms, the overview of the state-of-the-art deep learning models addressing BCI challenges, as well as the advanced BCI applications.

The rest of this book is organized as follows. We first propose a new taxonomy of BCI systems based on the category of brain signals in Chapter 2, followed by the basics of deep learning models including the concepts, architectures, and brief mathematical theories (Chapter 3). We provide a high-level overview of the frontiers of deep learning-based BCI studies (Chapter 4) and applications (Chapter 5). In addition, we present a set of advanced deep learning algorithms and frameworks aiming at the faced BCI challenges including robust representation learning (Chapter 6), cross-scenario classification (Chapter 7), and semi-supervised learning (Chapter 8). Furthermore, we provide the case studies of several novel BCI applications such as person authentication (Chapter 9), visual reconstruction (Chapter 10), language interpretation (Chapter 11), assistive living (Chapter 12), and neurological disorder diagnosis (Chapter 13). At last, Chapter 14 points out the future directions.

In terms of practical resources, we provide a detailed tutorial introducing every step of deep learning-based BCI along with implemantable codes in

https://github.com/xiangzhang1015/Deep-Learning-for-BCI

that even beginners in BCI or deep learning can understand and play around. All the necessary codes and data sets used in this book have been open-sourced for reproduction, please refer to the above link. Moreover, we present 31 reusable brain signal data sets in Section 5.9 which covers most existing brain signal categories. Furthermore, we also refer the interested readers to some public deep learning codes[2] for typical EEG signals and a toolbox[3] called neuro-imaging.

---

[2]https://github.com/NeuroTechX/dl-eeg-playground. Accessed Jan. 7, 2019.
[3]https://github.com/QTIM-Lab/DeepNeuro. Accessed Jun. 24, 2019.

Chapter 2

Brain Signal Acquisition

In this chapter, a comprehensive and systematic introduction of brain signals used in brain–computer interface (BCI) systems is presented. Figure 2.1 shows a new taxonomy of brain signals including invasive and noninvasive signals based on the signal collection method. Invasive signals are collected from the cortex surface or under the cortex surface (Section 2.1); noninvasive signals are collected by the external sensors (Section 2.2). Electroencephalogram (EEG) plays a dominant role among noninvasive signals. Therefore, the EEG signal and its subordinate categories in specific are introduced in Section 2.3. The basic characteristics of various brain signals are summarized in Table 2.1.

2.1 Invasive Approaches

Here we only briefly introduce the basic knowledge about invasive methods as this book mainly focuses on noninvasive approaches. Invasive recordings are acquired by electrodes deployed under the scalp. Figure 2.2 (Leuthardt *et al.*, 2009) shows both "intraparenchymal signals" gathered from the cortex and "electrocorticography (ECoG)"[1] gathered from the surface of the cortex (dura and arachnoid).

Invasive techniques can provide high-quality brain signals as electrodes collect signals directly from locations near the brain neurons. The collected

---

[1]Some studies refer to intracortical as "invasive," and ECoG as "semi-invasive." In this book, we combine the so-called "invasive" and "semi-invasive" into "invasive" because they both require surgery.

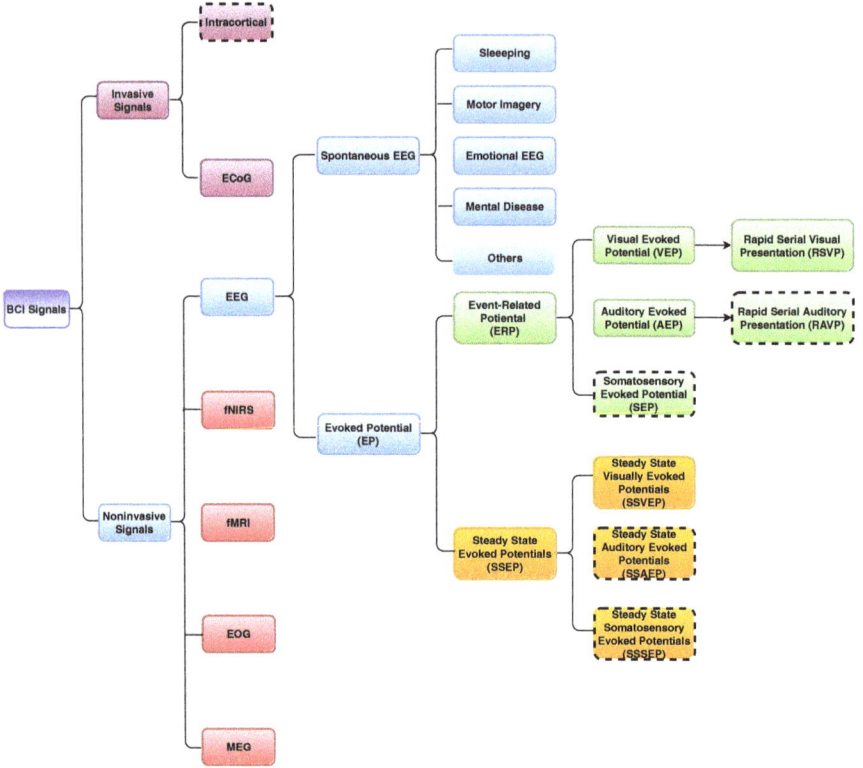

Fig. 2.1: The brain signal acquisition methods. The dashed quadrilaterals (intracortical, RAVP, SEP, SSAEP, and SSSEP) are not included in this chapter because there is no existing work focusing on them involving deep learning algorithms. P300, which is a positive potential recorded approximately 300 ms after the onset of presented stimuli, is not listed in this signal tree because it is included by ERP (which refers to all the potentials after the presented stimuli). In this classification, other brain signals beyond EEG (e.g., MEG and fNIRS) could also include visual/auditory tasks theoretically, but we omit them since there is no existing work adopting deep learning on these tasks.

signals have high temporal and spatial resolution[2] and high signal-to-noise ratio (SNR). Nevertheless, invasive methods suffer from two challenges. First, the implantation of electrodes requires a surgical procedure, which is expensive and risky because of potential medical complications such as transplant rejection. Second, implanted electrodes are fixed and, therefore,

---

[2]Spatial resolution refers to how well the signal discriminates among nearby locations.

Table 2.1: Summary of the characteristics of brain signals.

Signals	Invasive		Noninvasive				
	Intracortical	EcoG	EEG	fNIRS	fMRI	EOG	MEG
Invasiveness	High	High	Low	Low	Low	Low	Low
Spatial Resolution	Very high	High	Low	Intermediate	High	Low	Intermediate
Temporal Resolution	High	High	High	Low	Low	Intermediate	High
Signal-to-Noise Ratio	High	High	Low	Low	Intermediate	Intermediate	Low
Portability	Intermediate	Intermediate	High	High	Low	High	Low
Cost	High	High	Low	Low	High	Low	High
Characteristic	Electrical	Electrical	Electrical	Metabolic	Metabolic	Electrical	Magnetic

Layers   Signal Source

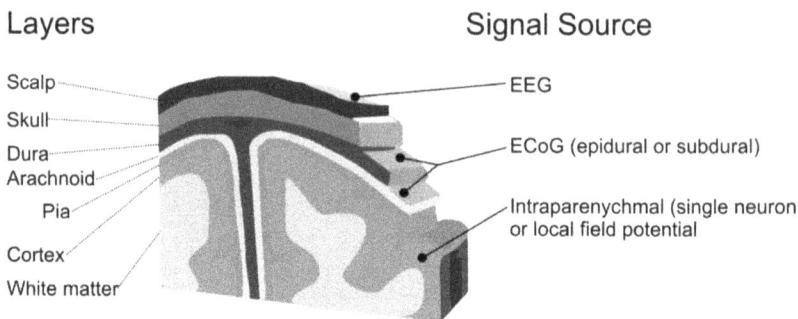

Scalp ·········
Skull ···········
Dura ···········
Arachnoid ···
    Pia ···
Cortex ···
White matter

EEG
ECoG (epidural or subdural)
Intraparenychmal (single neuron or local field potential

Fig. 2.2: Signal source locations in the brain (Leuthardt *et al.*, 2009).

can only measure the brain signals from the same locations. For these reasons, invasive brain signal techniques are mainly used in animals (e.g., monkeys and rats) and for people with severe disabilities (e.g., individuals with amyotrophic lateral sclerosis (ALS)) (Abdulkader *et al.*, 2015).

2.1.1   *Intracortical Approaches*

The intracortical recording technique involves the insertion of electrodes into the cortex of the subject's brain (Figure 2.2). The implanted micro-electrode can be a single electrode or an array of electrodes. Generally, the intracortical electrodes provide high-resolution motor control brain signals, as movement is the most easily observable phenomenon compared to other phenomena, such as hearing. Under the cortex, the electrodes are sensitive enough to pick up the discrete all-or-none output of single neurons, the action potential, commonly referred to as a "spike", as well as the summed voltage fluctuations from small to large numbers of neurons, called field potentials. Each electrode provides spiking from up to a few neurons, yielding the time-evolving output pattern of the population. These represent but a small sample of the entire set of neurons in this limited region, as spiking can only be detected by microelectrodes closely approximated to a neuron (Homer *et al.*, 2013). Pandarinath *et al.* (2017) developed a high-performance BCI system for communication among patients with ALS. This work implanted a 96-channel silicon microelectrode array in the motor cortex corresponding to hand area and recorded the motor intention of users by the microelectrode array. The array was then decoded into point-and-click commands to control a cursor.

2.1.2 *Electrocorticography*

ECoG is an extracortical invasive electrophysiological monitoring method to record brain activity. The electrodes collecting ECoG are attached under the skull, above (epidural) or below (subdural) the dura mater, but not within the brain parenchyma itself (Figure 2.2) (Leuthardt *et al.*, 2009). ECoG provides a tradeoff between higher SNR compared to noninvasive recordings and lower risk compared to intracortical recordings. It provides a higher spatial resolution and a rather high SNR with a lower surgical risk. Therefore, ECoG has a better prospect in the medical arena than intracortical recordings.

The ECoG collection approach and signals are shown in Figure 2.3 (Bandt *et al.*, 2017). ECoG signals have a higher amplitude compared to noninvasive brain wave signals. For instance, ECoG amplitude is higher than 50 μV, whereas the EEG amplitude is generally lower than 20 μV. The higher amplitude renders ECoG less vulnerable to artifacts such as eye blink activity. Moreover, ECoG has a bandwidth of 0–500 Hz, which is much wider than EEG (0–40Hz), because of the low-pass filtering effects of the skull. The wider frequency bands take substantial information from functional areas of a brain (e.g., motor and language) and thus can be used to train a higher performance BCI system. However, the disadvantages of an invasive methods like ECoG (such as the risky surgery and inconvenience of permanently attached devices) naturally limit its wide deployment in real-world scenarios.

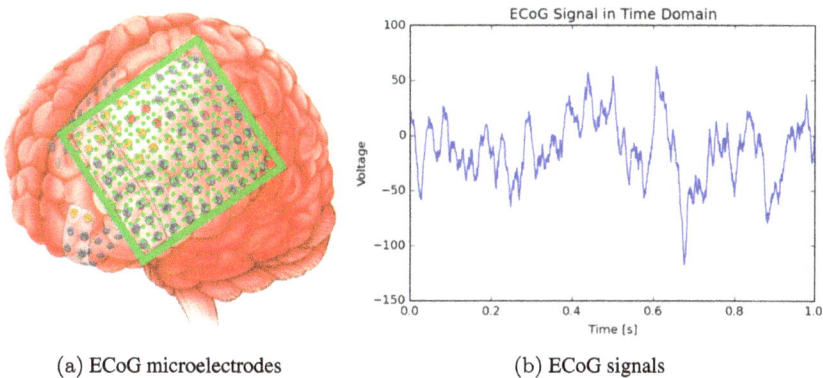

(a) ECoG microelectrodes (b) ECoG signals

Fig. 2.3: ECoG grid on cortical surface and ECoG signals (Bandt *et al.*, 2017).

2.2   Noninvasive Approaches

Noninvasive recordings can gather user's brain information without electrodes being inserted. Signals can be collected using electrical, magnetic, or metabolic methods. Noninvasive signals mainly include EEG, functional near-infrared spectroscopy (fNIRS), functional magnetic resonance imaging (fMRI), electrooculography (EOG), and magnetoencephalography (MEG). EEG-related studies represent the considerable majority of noninvasive signals and have numerous subclasses. We will introduce more details and subcategories of EEG in Section 2.3.

2.2.1   *Electroencephalography*

EEG is the most commonly used noninvasive technique for measuring brain activities. EEG monitors the voltage fluctuations generated by an electrical current within human neurons. Electrodes placed on the scalp measure the amplitude of EEG signals. EEG signals have a low spatial resolution because the number of electrodes is limited. EEG electrode locations generally follow the international 10–20 system or the intermediate 10% electrode positions (Malmivuo and Plonsey, 1995). The international 10–20 system divides the scalp in 10% and 20% intervals and totally contains 21 electrode locations (Figure 2.4). The intermediate 10% electrode position is standardized by the American Electroencephalographic Society and splits the scalp with 10% intervals, containing 75 electrodes. The existing EEG collection system is generally less than 75 electrodes, specifically, 64 electrodes (BCI 2000 system), 32 electrodes (openBCI headset), 14 electrodes (Emotiv EPOC+ headset), 5 electrodes (Emotiv insight headset), and 1 electrode (Mindware headset).

The temporal resolution of EEG signals is much better than spatial resolution. The ionic current changes rapidly, which offers a temporal resolution higher than 1000 Hz. The SNR of EEG is generally very poor because of both objective and subjective factors. Objective factors include environmental noises, the obstruction of the skull and other tissues between cortex and scalp, and various stimulations. Subjective factors contain the mental stage of the subject, fatigue status, and the variance among different subjects.

EEG-recording equipment can be installed in a cap-like headset. As shown in Figure 2.5 (Zhang *et al.*, 2018h), the EEG headset can be mounted on the head of the user to gather signals. Compared to other

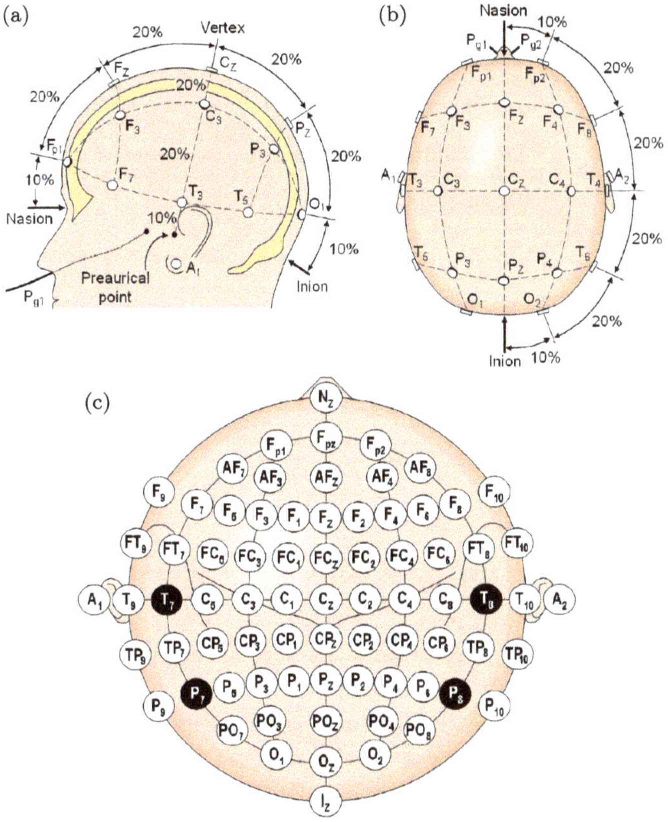

Fig. 2.4: (a) and (b) are the left and above view of the international 10–20 system; (c) presents the intermediate 10% electrodes positions (Sazgar and Young, 2019).

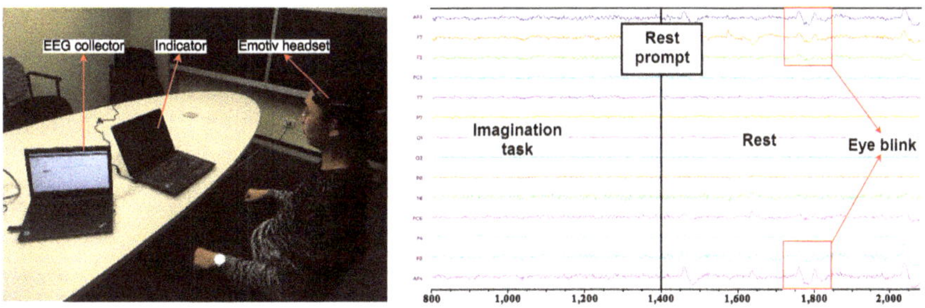

(a) EEG

(b) EEG signals

Fig. 2.5: EEG collection scenario and the gathered signals. The subject is undertaking an imagination task.

Table 2.2: EEG patterns and corresponding characters. Awareness degree denotes the degree of being aware of an external world. Consciousness represents the subject's normal state of being awake.

Patterns	*Hz*	Amplitude	Brain State	Awareness	Produced Location
Delta	0.5–4	Higher	Deep sleep pattern	Lower	Frontally and posteriorly
Theta	4–8	High	Light sleep pattern	Low	Entorhinal cortex, hippocampus
Alpha	8–12	Medium	Closing the eyes, relax state	Intermediate	Posterior regions of head
Beta	12–30	Low	Active thinking, focus	High	Most evident frontally
Gamma	30–100	Lower	Cross-modal sensory processing	Higher	Somatosensory cortex

equipment used to measure brain signals, EEG headsets are portable and more accessible for most applications.

The EEG signals collected from any typical EEG hardware have several nonoverlapping frequency bands (Delta, Theta, Alpha, Beta, and Gamma) based on the strong intraband correlation with a distinct behavioral state (Zhang *et al.*, 2018h). Each EEG pattern contains signals associated with particular brain information. Table 2.2 shows EEG frequency patterns and the corresponding characteristics. The degree of awareness denotes the perception of individuals when presented with external stimuli. It is mainly defined in physiology instead of psychology. Each frequency band represents a brain state and a qualitative assessment of awareness:

- Delta pattern (0.5–4 Hz) corresponds to deep sleep when the subject has lower awareness.
- Theta pattern (4–8 Hz) corresponds to light sleep in the realm of low awareness.
- Alpha pattern (8–12 Hz) mainly occurs during eyes closed and deeply relaxed state and corresponds to the medium awareness.
- Beta pattern (12–30 Hz) is the dominant rhythm while the eyes of the subject are open and is associated with high awareness. Beta patterns capture most of our daily activities (such as eating, walking, and talking).

- Gamma pattern (30–100 Hz) represents the co-interaction of several brain areas to carry out a specific motor and cognitive function.

2.2.2 *Functional Near-infrared Spectroscopy*

fNIRS is a noninvasive functional neuroimaging technology using near-infrared (NIR) light (Naseer *et al.*, 2016). In specific, fNIRS employs NIR light to measure the aggregation degree of oxygenated hemoglobin (Hb) and deoxygenated-hemoglobin (deoxy-Hb) because Hb and deoxy-Hb have higher absorbance of light than other head components such as the skull and scalp. fNIRS relies on blood-oxygen-level-dependent (BOLD) response or hemodynamic response to form a functional neuroimage. The BOLD response can detect the oxygenated or deoxygenated blood level in the brain blood. The relative levels reflect the blood flow and neural activation, where increased blood flow implies a higher metabolic demand caused by active neurons. For example, when the user is concentrating on a mental task, the prefrontal cortex neurons will be activated, and the BOLD response in the prefrontal cortex area will be stronger (Hennrich *et al.*, 2015).

Figure 2.6 shows the fNIRS collection hardware and the collected signals. Single or multiple emitter–detector pairs measure the Hb and deoxy-Hb: the emitter transmits NIR light through the blood vessels to the detector. Most existing studies use fNIRS technologies to measure

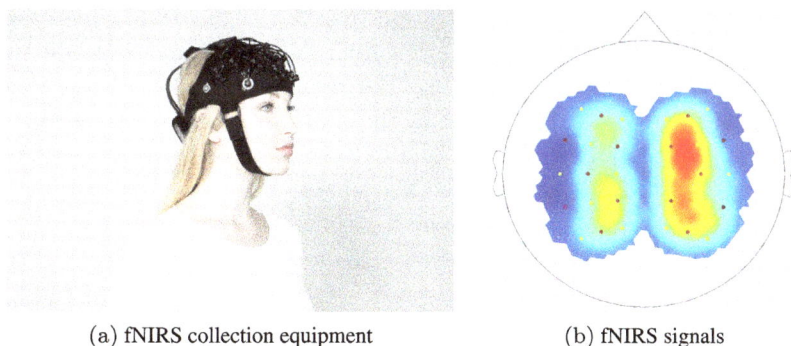

(a) fNIRS collection equipment   (b) fNIRS signals

Fig. 2.6: fNIRS collection equipment and the gathered signals.[3]

---

[3]https://www.artinis.com/fnirs. Accessed Jan. 20, 2017.

the status of prefrontal and motor cortex. The former is a response to mental tasks, whereas the latter is a response to motor-related tasks (e.g., motor imagery). The monitored Hb and deoxy-Hb change slowly since the blood speed varies in a relatively slow ratio compared to electrical signals. Therefore, fNIRS signals have lower temporal resolution[4] compared with electrical or magnetic signals. The spatial resolution depends on the number of emitter–detector pairs. In current studies, three emitters and eight detectors would suffice for adequately acquiring the prefrontal cortex signals; and six emitters and six detectors would suffice for covering the motor cortex area (Naseer and Hong, 2015). A drawback of fNIRS is that it cannot be used to measure cortical activity occurring deeper than 4 cm in the brain, because of the limitations in light emitter power and spatial resolution.

2.2.3 *Functional Magnetic Resonance Imaging*

fMRI monitors brain activities by detecting changes associated with blood flow in brain areas (Wen *et al.*, 2018). Similar to fNIRS, fMRI relies on the BOLD response. The main differences between fNIRS and fMRI are as follows (Liu *et al.*, 2018a). First, as the name implies, fMRI measures BOLD response through magnetic instead of optical methods. Hemoglobin differs in how it responds to magnetic fields, depending on whether it has a bound oxygen molecule. The magnetic fields are more sensitive to and are more easily distorted by deoxy-Hb than Hb molecules. Second, the magnetic fields have higher penetration than NIR light, which gives fMRI greater ability to capture information from deep parts of the brain than fNIRS. Third, fMRI has a higher spatial resolution than fNIRS since the latter's spatial resolution is limited by the emitter–detector pairs. However, the temporal resolutions of fMRI and fNIRS are at an equal level because they both constrained by the blood flow speed.

fMRI has several flaws compared to fNIRS: (1) fMRI requires an expensive scanner to generate magnetic fields; (2) the scanner is heavy and has poor portability. Figure 2.7 shows the fMRI acquisition machine, and the resulting brain images. fMRI images of speech perception and finger tapping have a significant difference, indicating high SNR.

---

[4]Temporal resolution refers to the smallest time period of neural activity reliably separated out.

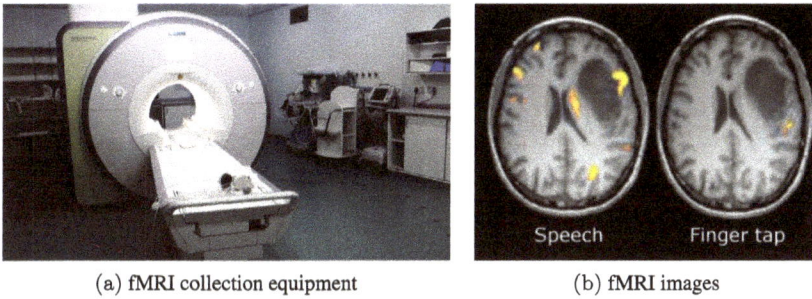

(a) fMRI collection equipment (b) fMRI images

Fig. 2.7: fMRI collection equipment and the gathered fMRI signals while the subject is speaking and finger tapping.[5]

2.2.4 *Electrooculography*

EOG is a technique for measuring the corneo-retinal standing potential that exists between the front and the back of the human eyes. Most patients who have lost voluntary motor movements (e.g., locked-in syndrome patients) remain in partial control of the eyes (Yang *et al.*, 1989). The eye movements can be detected by EOG signals to interact with the external devices. Therefore, we regard EOG signals as one class of brain signals in this book. EOG can be used to bridge the user and the outer world because different eye movements will cause different electrical potentials. Pairs of electrodes are typically placed above/below the eye or to the left/right of the eye to measure EOG signals. The EOG collection equipment (Aungsakun *et al.*, 2011) and the collected signals (Rusydi *et al.*, 2014) can be found in Figure 2.8. Figure 2.8(a) shows EOG electrode placements, where electrodes Ch.V+ and Ch.V– measure the vertical movements, and Ch.H+ and Ch.H– measure the horizontal movements. G electrode representing the ground line works as a reference point. Figure 2.8(b) shows the vertical EOG in the time domain under six scenarios (looking upward, looking downward, single blink, double blink, looking leftward, and looking rightward). We can observe EOG signals have large variances among different scenarios, indicating they have a relatively high SNR and are easily recognizable by machine learning algorithms. EOG has low spatial resolution compared to other brain signals since we can

---

[5]https://www.jameco.com/Jameco/workshop/HowItWorks/what-is-an-fmri-scan-and-how-does-it-work.html. Accessed Jan. 5, 2017.

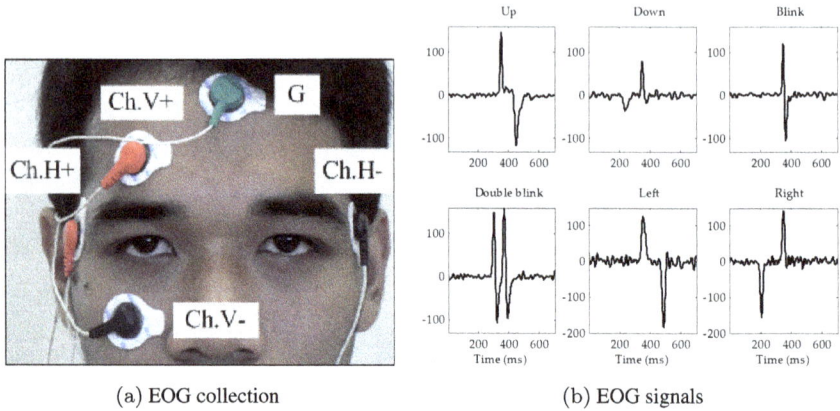

(a) EOG collection　　　　　　　　(b) EOG signals

Fig. 2.8: EOG collection equipment (Aungsakun *et al.*, 2011) and the gathered vertical signals while the subject is looking in different directions and blinking (Rusydi *et al.*, 2014).

only detect the vertical and horizontal potentials. The temporal resolution of EOG is higher than neuroimaging techniques because the electrical potentials vary faster than metabolic features (e.g., blood flow).

2.2.5　*Magnetoencephalography*

MEG is a functional neuroimaging technique for mapping brain activity by recording magnetic fields produced by electrical currents occurring naturally in the brain, using very sensitive magnetometers (Cichy *et al.*, 2017). The ionic currents of active neurons will create weak magnetic fields. The generated magnetic fields can be measured by magnetometers like SQUIDs (superconducting quantum interference devices). However, producing a detectable magnetic field requires massive (e.g., 50,000) active neurons with similar orientation. The source of the magnetic field measured by MEG is the pyramidal cells which are perpendicular to the cortex surface.

MEG has a relatively low spatial resolution because the signal quality highly depends on the measurement factors (e.g., brain area, neuron orientations, neuron depth). However, MEG can provide very high temporal resolution ($\geq 1000\,\text{Hz}$) because MEG directly monitors the brain activity from the neuron level, which is in the same level of intracortical signals. The MEG equipment (Ukil, 2006) is shown in Figure 2.9. MEG equipment is expensive and not portable, which limits its real-world

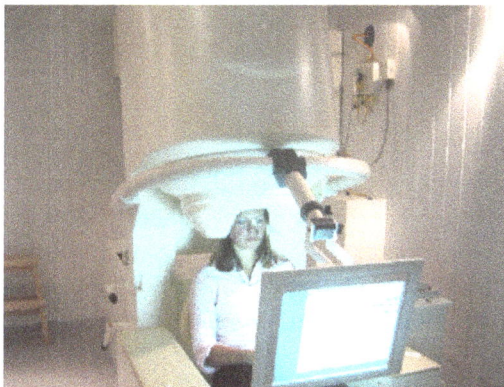

Fig. 2.9: MEG collection equipment (Ukil, 2006).

deployment for brain signal. The gathered MEG signal is similar to EEG signals.

2.3 EEG Paradigms

Compared to other noninvasive signals (e.g., fMRI, fNIRS, EOG, MEG), EEG has several important advantages: (1) the hardware has higher portability with much lower price; (2) the temporal resolution is very high (milliseconds level)[6]; (3) EEG is relatively tolerant of subject movement and artifacts, which can be minimized by existing signal processing methods; (4) the subject doesn't need to be exposed to high-intensity (>1 Tesla) magnetic fields. Thus, EEG can serve subjects that have metal implants in their body (such as metal-containing pacemakers).

As the most commonly used signals, there are a huge number of subcategories of EEG signals. In this section, we present a systematic introduction of EEG subclass signals. As shown in Figure 2.1, we divided EEG signals into spontaneous EEG and evoked potentials (EPs). EPs can be split into event-related potentials (ERPs) and steady-state evoked potentials (SSEP) based on the frequency of the external stimuli. Each potential contains visual-, auditory-, and somatosensory-potentials based on the external stimuli types. The dashed quadrilaterals in Figure 2.1, such as intracortical, SEP, SSAEP, SSSEP, and RSAP, are not included in this

---

[6]http://www.biomagcentral.org/biomagnetism/meg.html. Accessed Feb. 23, 2020.

book because there are very few existing studies working on them with deep learning algorithms. We list these signals for systematic completeness.

2.3.1 *Spontaneous EEG*

Generally, when we talk about the term "EEG," we refer to *spontaneous EEG*, which measures the brain signals under a specific state without external stimulation. For example, spontaneous EEG includes the EEG signals while the user is sleeping, undertaking a mental task (e.g., counting), under fatigue stage, suffering brain disorders, undertaking motor imagery tasks, and so on.

The EEG signals recorded while a user stares at a color/shape/image belong to this category. While the subject is gazing at a specific image, the visual stimuli are steady without any change. This scenario differs from the visual stimuli in EP, where the visual stimuli are changing at a specific frequency. Thus, we regard the image stimulation as a particular state and categorize it as spontaneous EEG. BCI systems based on spontaneous EEG are challenging to train, because of the lower SNR and the larger variation across subjects (Pfurtscheller and Neuper, 2001).

2.3.2 *Evoked Potential*

EP or evoked responses refer to EEG signals which are evoked by a event stimulus instead of spontaneously. An EP is time-locked to the external stimulus, whereas the aforementioned spontaneous EEG is non-time-locked. In contrast to spontaneous EEG, EP generally has higher amplitude and lower frequency. As a result, the EP signals are more robust across subjects. According to the stimulation method, there exist two categories of EP: the ERP and the SSEP. ERP records the EEG signals in response to an isolated discrete stimulus event. To achieve this isolation, stimuli in an ERP experiment are typically separated from each other by a long inter-stimulus interval, allowing for the estimation of a stimulus-independent baseline reference (Norcia *et al.*, 2015). The stimuli frequency of ERP is generally lower than 2 Hz. In contrast, SSEP is generated in response to a periodic stimulus at a fixed rate. The stimuli frequency of SSEP generally ranges within 3.5–75 Hz.

2.3.2.1 *Event-Related Potential*

There are three kinds of EPs in extensive research and clinical use: visual evoked potentials (VEP); auditory evoked potentials (AEP); and somatosensory evoked potentials (SEP) (Cecotti and Ries, 2017). The VEP signals are mainly on the occipital lobe, and the highest signal amplitudes are collected at the calcarine sulcus.

(1) VEP. VEPs are a specific category of ERP caused by visual stimulus (e.g., an alternating checkerboard pattern on a computer screen). VEP signals are hidden within the normal spontaneous EEG. To separate VEP signals from the background EEG readings, repetitive stimulation and time-locked signal-averaging techniques are generally employed.

Rapid serial visual presentation (RSVP) (Lees *et al.*, 2018) can be regarded as one kind of VEP. An RSVP diagram is commonly used to examine the temporal characteristics of attention. The subject is required to stare at a screen where a series of items (e.g., images) are presented one by one. There is a specific item (called the target) that separates from the rest of the other items (called distracters). The subject knows which is the target before the RSVP experiment. Generally, the distracters can either be a color change or letters among numbers. RSVP contains a static mode (the items appear on the screen and then disappear without moving) and a moving mode (the items appear on the screen, move to another place, and finally disappear). Nowadays, brain signal research mainly focuses on the static mode RSVP. Usually, the frequency of RSVP is 10 Hz which means that each item will stay on the screen for 0.1 s.

(2) AEP. AEPs are a specific subclass of ERP in which responses to auditory (sound) stimuli are recorded. AEP is mainly recorded from the scalp but originates at the brainstem or cortex. The most common AEP measured is the auditory brainstem response that is often employed to test the hearing ability of newborns and infants. In BCI, AEP is mainly used in clinical tests for its accuracy and reliability in detecting unilateral loss (Chiappa, 1997). Similar to RSVP, rapid serial auditory presentation (RSAP) refers to experiments with rapid serial presentation of sound stimuli. The task for the subject is to recognize the target audio among the distractors.

(3) SEP.[7] SEP is another commonly used subcategory of ERP, which is elicited by electrical stimulation of the peripheral nerves. SEP signals include a series of amplitude deflection that can be elicited by virtually any sensory stimuli.

---

[7]Generally, SEPs are abbreviated as SSEP or SEP. In this chapter, we choose SEP as the abbreviation in case of conflict with SSEPs.

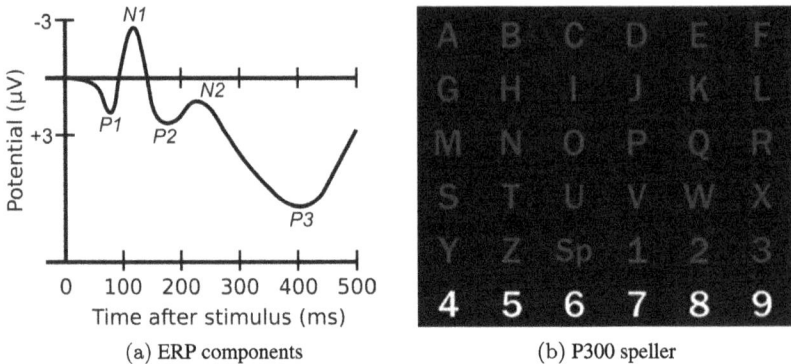

(a) ERP components

(b) P300 speller

Fig. 2.10: P300 waves and visual-based P300 speller (Farwell and Donchin, 1988).

2.3.2.2 *P300*

P300 (also called P3) is an important component in ERP (Guger *et al.*, 2009). Here we introduce P300 signal separately since it is widely used for BCI systems. Figure 2.10(a) shows the ERP signal fluctuation in the 500 ms after the stimuli onset.[8] The waveform mainly contains five components, P1, N1, P2, N2, and P3. The capital character P/N represents positive/negative electrical potentials. The following number refers to the occurrence time of the specific potential. Thus, P300 denotes the positive potential of ERP waveform at approximately 300 ms after the presented stimuli. Compared to other components, P300 has the highest amplitude and is easiest to detect. Thus, a large number of brain signal studies focus on P300 analysis. P300 is more of an informative feature instead of a type of brain signal (e.g., VEP). Therefore, we do not list P300 in Figure 2.1. P300 can be analyzed in most of ERP signals such as VEP, AEP, SEP.

In practice, P300 can be elicited by rare, task-relevant events in an "oddball" paradigm (e.g., P300 speaker). In the oddball paradigm, the subject receives a series of stimuli where low-probability target items are mixed with high-probability nontarget items. Visual and auditory stimuli are the most commonly used in the oddball paradigm. Figure 2.10(b) shows an example of visual-based P300 speller, which enables the subject to spell letters/numbers directly through brain signals (Farwell and Donchin, 1988). The 26 letters of the alphabet and the Arabic numbers are displayed

---

[8]The negative voltage of ERP is plotted upward, which is common in ERP research.

on a computer screen which serves as the keyboard. The subject focuses attention successively on the characters they wish to spell. The computer detects the chosen character online in real time. This detection is achieved by repeatedly flashing rows and columns of the matrix. When the elements containing the selected characters are flashing, a P300 fluctuation is elicited. In the 6×6 matrix screen, the rows and columns flash in mixed random order. The flash duration and interval among adjacent flashes are generally set as 100 ms (Belitski *et al.*, 2011). The columns and rows flash separately. First, the columns flash six times with each column flashing one time. Second, the rows will flash for six times. After that, this paradigm repeats for several times (e.g., N times). The P300 signals of the total $12N$ flash will be analyzed to output a single outcome (i.e., one letter/number).

2.3.2.3 *Steady State Evoked Potentials*

SSEPs are another subcategory of EPs, which are periodic cortical responses evoked by certain repetitive stimuli with a constant frequency. It has been demonstrated that the brain oscillations generally maintain a steady level over time, whereas the potentials are evoked by steady state stimuli (e.g., a flickering light with fixed frequency). Technically, SSEP is defined as a form of response to repetitive sensory stimulation in which the constituent frequency components of the response remain constant over time in both amplitude and phase (Regan, 1977). Depending on the type of stimuli, SSEP can be divided into three subcategories: steady-state visually evoked potentials (SSVEP), steady-state auditory evoked potentials (SSAEP), and steady-state somatosensory evoked potentials (SSSEP). In the brain signal area, most studies are focused on visual evoked steady potentials, and papers only rarely focus on auditory and somatosensory stimuli. Therefore, in this book, we mainly introduce SSVEP rather than SSAEP and SSSEP.

2.3.2.4 *Commonly Used Visual-Related Potentials*

VEPs are the most commonly used potentials. Therefore, it is essential to distinguish the three different VEP paradigms: VEP, RSVP, and SSVEP. Here, we theoretically introduce the characteristics of each paradigm. First, the frequencies are different: the frequency of VEP is less than 2 Hz, whereas the frequency of RSVP is around 10 Hz, and the frequency of SSVEP ranges from 3.5 to 75 Hz. In addition, they have various presentation protocols. In the VEP paradigm, different visual patterns will be presented on the screen *in turn* to check the changes in the brain signals of the user. In an RSVP diagram, several items will be presented on a screen *one by one*. All the items are shown in the same place and share the same frequency. In SSVEP paradigm, several items will be presented on a screen *at the same time* while the items are shown at *variant positions* with different frequencies.

Chapter 3

Deep Learning Foundations

In this chapter, the foundations of deep learning models including concepts, architectures, and techniques commonly used in the brain signal field will be formally introduced. Deep learning is a class of machine learning technique that uses many layers of information-processing stages in hierarchical architectures for pattern classification and feature/representation learning (Deng, 2014). A relatively detailed introduction of various deep learning models will be given for the reason that some of the potential readers who are from noncomputer backgrounds (e.g., biomedicine) will not be familiar with deep learning.

Deep learning algorithms are divided into several subcategories based on the aim of the techniques (as shown in Figure 3.1):

- Discriminative deep learning models, which classify the input data into a preknown label based on the adaptively learned discriminative features. Discriminative algorithms are able to learn distinctive features by nonlinear transformation, and classification through probabilistic prediction.[1] Thus these algorithms can play the role of both feature engineering and classification (corresponding to Figure 1.1). Discriminative

---

[1] The classification function is achieved by the combination of a softmax layer and one-hot label encoding. The one-hot label encoding refers to encoding the label by the one-hot method, which is a group of bits among which the only valid combinations of values are those with a single high (1) bit and all the others with low (0) bits. For instance, a set of labels 0, 1, 2, 3 can be encoded as (1, 0, 0, 0), (0, 1, 0, 0), (0, 0, 1, 0), (0, 0, 0, 1).

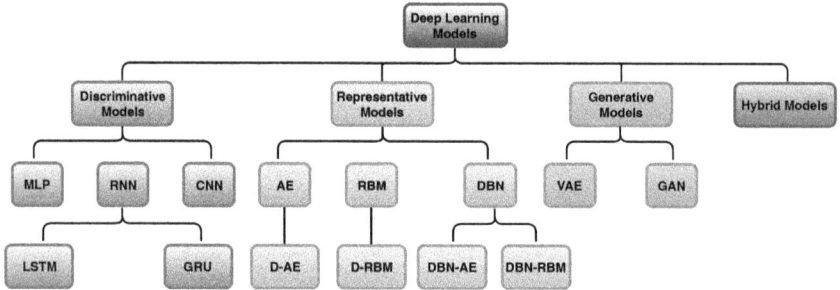

Fig. 3.1: Deep learning models. They can be divided into discriminative, representative, generative and hybrid models based on the algorithm function. D-AE denotes Stacked-Autoencoder which refers to the autoencoder (AE) with multiple hidden layers. Deep Belief Network (DBN) can be composed of AE or restricted Boltzmann machine (RBM), therefore, we divided DBN into DBN-AE (stacked AE) and DBN-RBM (stacked RBM).

architectures mainly include Multilayer Perceptron (MLP), Recurrent Neural Networks (RNN), Convolutional Neural Networks (CNN), along with their variations.

- Representative deep learning models, which learn the pure and representative features from the input data. These algorithms only have the function of feature engineering (corresponding to Figure 1.1) but fail to classify. Commonly used deep learning algorithms for representation are autoencoder (AE), restricted Boltzmann machine (RBM), deep belief networks (DBN), along with their variations.

- Generative deep learning models, which learn the joint probability distribution of the input data and the target label. In the brain signal scope, generative algorithms are mostly used to reconstruct or generate a batch of brain signal samples to enhance the training set. Generative models commonly used for brain signals, which include variational autoencoder (VAE),[2] generative adversarial networks (GANs), and so on.

- Hybrid deep learning models, which combine more than two deep learning models. For example, the typical hybrid deep learning model employs a representation algorithm for feature extraction and discriminative algorithms for classification.

The characteristics of each deep learning subcategories are summarized in Table 3.1. Almost all the classification functions in neural networks

---

[2]VAE is a variation of AE, but working on a different aspect. Therefore, AE and VAE are separately introduced.

Table 3.1: Summary of deep learning model types.

Deep Learning	Input	Output	Function	Training Method
Discriminative	Input data	Label	Feature extraction, classification	Supervised
Representative	Input data	Representation	Feature extraction	Unsupervised
Generative	Input data	New sample	Generation, reconstruction	Unsupervised
Hybrid	Input data	—	—	—

are implemented by a softmax layer, which will not be regarded as an algorithmic component. For instance, a model combining DBN and a softmax layer will still be regarded as a representative model instead of a hybrid model.

3.1 Discriminative Deep Learning Models

As the main task of brain signal research is brain signal recognition, discriminative deep learning models are the most popular and powerful algorithms. Suppose we have a data set of brain signal samples $\{\mathbb{X}, \mathbb{Y}\}$, where $\mathbb{X}$ denotes the set of brain signal observations and $\mathbb{Y}$ denotes the set of sample ground truth (i.e., labels). Suppose a specific sample–label pair is given by $\{\boldsymbol{x} \in \mathbb{R}^N, \boldsymbol{y} \in \mathbb{R}^M\}$, where N and M denote the dimension of observations and the number of sample categories, respectively. The aim of discriminative deep learning models is to learn a function with the mapping: $\boldsymbol{x} \to \boldsymbol{y}$. In short, the discriminative models receive the input data and output the corresponding category or label. All the discriminative models introduced in this section are supervised learning techniques, which require the information of both the observations and the ground truth.

3.1.1 *Multilayer Perceptron*

MLP, one of the simplest and the most basic deep learning models, is modified based on the standard neural network (Figure 3.2(a)), which contains three neuron layers (i.e., an input layer, a hidden layer, and an output layer). The key difference between MLP and the standard neural network is that MLP has more than one hidden layers. All the nodes are

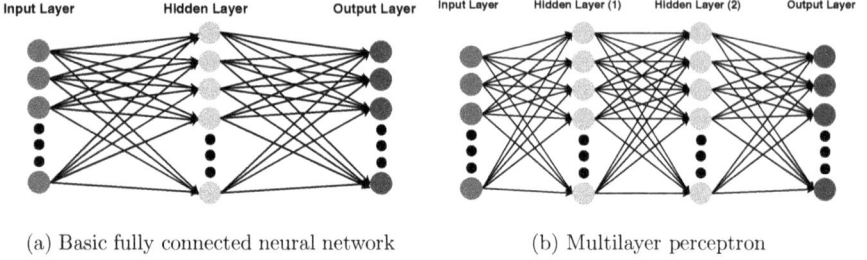

(a) Basic fully connected neural network

(b) Multilayer perceptron

Fig. 3.2: Illustration of standard neural network and MLP. (a) The basic structure of the fully connected neural network. The input layer receives the raw data or extracted features of brain signals while the output layer shows the classification results. The term "fully connected" denotes that each node in a specific layer is connected with all the nodes in the previous and next layers. (b) MLP could have multiple hidden layers, the more, the deeper. This is an example of MLP with two hidden layers, which is the simplest MLP model.

fully connected with the nodes of the adjacent layers but without connection with the other nodes of the same layer. MLP includes multiple hidden layers. As shown in Figure 3.2(b), we take a structure with two hidden layers as an example to describe the data flow in MLP. First, we define an operation $\mathcal{T}(\cdot)$ as

$$\mathcal{T}(\boldsymbol{x}) = \boldsymbol{w} * \boldsymbol{x} + \boldsymbol{b} \tag{3.1}$$

$$\mathcal{T}(\boldsymbol{x}, \boldsymbol{x}') = \boldsymbol{w} * \boldsymbol{x} + \boldsymbol{b} + \boldsymbol{w}' * \boldsymbol{x}' + \boldsymbol{b}' \tag{3.2}$$

where $\boldsymbol{x}$ and $\boldsymbol{x}'$ denote two variables, whereas $\boldsymbol{w}$, $\boldsymbol{w}'$, $\boldsymbol{b}$, and $\boldsymbol{b}'$ denote the corresponding weights and basis.

The input layer receives the observation $\boldsymbol{x}$ and feeds forward to the first hidden layer,

$$\boldsymbol{x}^{\boldsymbol{h1}} = \sigma(\mathcal{T}(\boldsymbol{x})) \tag{3.3}$$

where $\boldsymbol{x}^{\boldsymbol{h1}}$ denotes the data flow in the first hidden layer and σ represents the nonlinear activation function. There are several commonly used activation functions such as sigmoid/Logistic, Tanh, ReLU; sigmoid activation function is chosen as an example in this section. Then, the data flow to the second hidden layer and the output layer,

$$\boldsymbol{x}^{\boldsymbol{h2}} = \sigma(\mathcal{T}(\boldsymbol{x}^{\boldsymbol{h1}})) \tag{3.4}$$

$$\boldsymbol{y}' = \sigma(\mathcal{T}(\boldsymbol{x}^{\boldsymbol{h2}})) \tag{3.5}$$

where y' denotes the predict results in one-hot format. The error (i.e., loss) could be calculated based on the distance between y' and the ground truth y. For instance, the Euclidean-distance-based error can be calculated by

$$error = \|y' - y\|_2 \tag{3.6}$$

where $\|\cdot\|_2$ denotes the Euclidean norm. Afterward, the error will be back-propagated and optimized by a suitable optimizer. The optimizer will adjust all the weights and basis in the model until the error converges. The most widely used loss functions include cross-entropy, negative log likelihood, mean square estimation, and so on. The most widely used optimizers include adaptive moment estimation (Adam), stochastic gradient descent (SGD), adaptive sub-gradient method (Adagrad), and so on.

Several terms may be easily confused with each other: artificial neural network (ANN), deep neural network (DNN), and MLP. These terms have no strict difference and are often mixed in literature. Generally, ANN represents neural networks with fewer hidden layers (shallow), whereas DNN has more (in this case, DNN is equivalent to MLP).

3.1.2 *Recurrent Neural Networks*

RNN is a specific subcategory of discriminative deep learning model designed to capture temporal dependencies among input data. Figure 3.3(a) describes the activity of a specific RNN node in the time domain. At each time slice $t, t \in [1, t+1]$, the node receives an input I_t[3] and a hidden state c from the previous time (except the first time). For instance, at time t it receives not only the input I_t but also the hidden state of the previous node c_{t-1}. The hidden state can be regarded as the "memory" of the nodes which can help the RNN "remember" the historical input.

Next, we will report two typical RNN architectures that have attracted much attention and achieved great success: LSTM and GRUs. They both follow the basic principles of RNN, and we will pay attention to the complicated internal structures in each node. As the structure is much more complicated than general neural nodes, we call it a "cell." Cells in RNN are equivalent to nodes in MLP.

---

[3]The subscript represents the specific time.

(a) Recurrent Neural Networks

(b) Convolutional Neural Networks

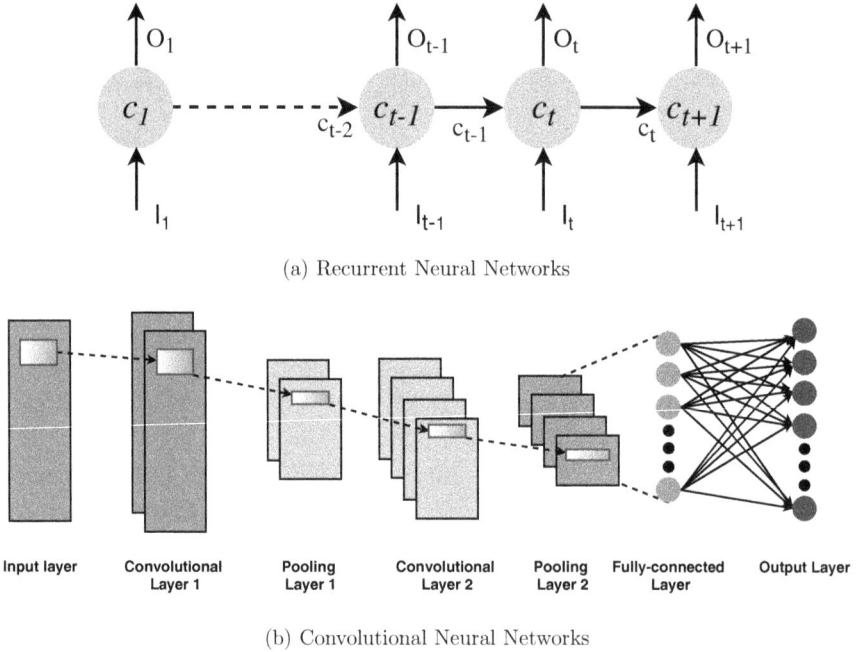

Fig. 3.3: Illustration of RNN and CNN models. (a) The recurrent procedure of the RNN model. This procedure describes the recurrent procedure of a specific node in time range $[1, t+1]$. The node at time t receives two input variables (I_t denotes the input at time t and c_{t-1} denotes the hidden state at time $t-1$) and exports two variables (the output O_t and the hidden state c_t at time t). (b) The paradigm of CNN model which includes two convolutional layers, two pooling layers, and one fully connected layer.

3.1.2.1   *Long Short-Term Memory*

Figure 3.4(a) shows the structure of a single LSTM cell at time t. The LSTM cell has three inputs (I_t, O_{t-1}, and c_{t-1}) and two outputs (c_t and O_t). The operation is as follows:

$$I_t, O_{t-1}, c_{t-1} \rightarrow c_t, O_t \tag{3.7}$$

I_t denotes the input value at time t, O_{t-1} denotes the output at the previous time (i.e., time $t-1$), and c_{t-1} denotes the hidden state at the previous time. c_t and O_t separately denote the hidden state and the output at time t. Therefore, we can observe that the output O_t at time t is not only related to the input I_t but also related to the information at the previous time. In this way, LSTM is empowered to remember important information in the

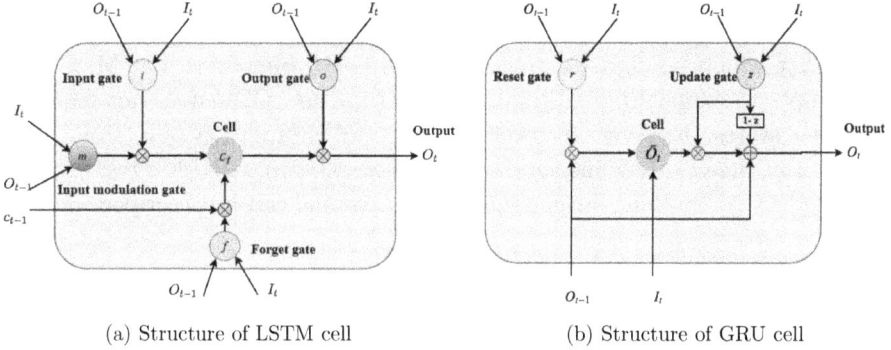

(a) Structure of LSTM cell   (b) Structure of GRU cell

Fig. 3.4: Illustration of detailed long short-term memory (LSTM) and gated ecurrent unit (GRU) cell structures. (a) LSTM cell receives three inputs (I_t denotes the input at time t, O_{t-1} denotes the output of previous time, and c_{t-1} denotes the hidden state of the previous time) and exports two outputs (the output of this time O_t and the hidden state of this time c_t). (b) GRU cell receives two inputs (the input of this time I_t and the output of the previous time O_{t-1}) and exports its output O_t. Unlike the hidden state c_t in LSTM cell, there is no transmittable hidden state in GRU cell except one intermediate variable $\bar{O}_t$.

time domain. Moreover, the essential idea of the LSTM is to control the memory of specific information. For this aim, the LSTM cell adopts four gates: the input gate, forget gate, output gate, and input modulation gate. Each gate is a weight to control how much information can flow through this gate. For example, if the weight of the forget gate is zero, the LSTM cell would remember all the information passed from the previous time $t-1$; if the weight is one, the LSTM cell would remember nothing. The corresponding activation function determines the weight. The detailed data flow is as follows:

$$f = \sigma(\mathcal{T}(I_t, O_{t-1})) \tag{3.8}$$

$$i = \sigma(\mathcal{T}(I_t, O_{t-1})) \tag{3.9}$$

$$o = \sigma(\mathcal{T}(I_t, O_{t-1})) \tag{3.10}$$

$$m = tanh(\mathcal{T}(I_t, O_{t-1})) \tag{3.11}$$

$$c_t = f * c_{t-1} + i * m \tag{3.12}$$

$$h_t = o * tanh(c_t) \tag{3.13}$$

where i, f, o, and m represent the input gate, forget gate, output gate, and input modulation gate, respectively.

3.1.2.2 *Gated Recurrent Units*

Another widely used RNN architecture is GRU. Similar to LSTM, GRU attempts to exploit the information from the past. GRU does not require hidden states; however, it receives temporal information only from the output of time $t-1$. Thus, as shown in Figure 3.4(b), GRU has two inputs (I_t and O_{t-1}) and one output (O_t). The mapping can be described as:

$$I_t, O_{t-1} \to O_t \tag{3.14}$$

GRU contains two gates: reset gate r and update gate z. The former decides how to combine the input with previous memory. The latter decides how much of its previous memory to keep around, which is similar to the forget gate of LSTM. The data flow is as follows:

$$z = \sigma(\mathcal{T}(I_t, O_{t-1})) \tag{3.15}$$

$$r = \sigma(\mathcal{T}(I_t, O_{t-1})) \tag{3.16}$$

$$\bar{O}_t = tanh(\mathcal{T}(I_t, r * O_{t-1})) \tag{3.17}$$

$$O_t = (1 - z) * O_{t-1} + z * \bar{O}_t \tag{3.18}$$

It can be observed that there's an intermediate variable $\bar{O}_t$ which is similar to the hidden state of LSTM. However, $\bar{O}_t$ only works on this time point and unable to pass to the next time point.

A brief comparison between LSTM and GRU is given as they are very similar. First, LSTM and GRU have comparable performance as studied by literature. For any specific task, it is recommended to try both of them to determine which provides better performance. Second, GRU is lightweight because it only has two gates and no hidden state. Therefore, GRU costs less training time and requires less data for generalization. In contrast, LSTM generally works better if the training data set is big enough.

3.1.3 *Convolutional Neural Networks*

CNN is one of the most popular deep learning models specialized in spatial information exploration. We will briefly present the working mechanism of CNN. CNN discovers the latent spatial information in applications such as image recognition, ubiquitous, and object searching due to their salient features such as regularized structure, good spatial locality, and translation invariance. In brain signal research, specifically, CNN is supposed to capture the distinctive dependencies among the patterns associated with different brain signals.

We present a standard CNN architecture as shown in Figure 3.3(b). The CNN contains one input layer, two convolutional layers with each followed by a pooling layer, one fully connected layer, and one output layer. The square patch in each layer shows the processing progress of a specific batch of input values. The key to CNN is to reduce the input data into a form that is easier to recognize, with as little information loss as possible. CNN has three stacked layers: the convolutional layer, pooling layer, and fully connected layer.

The convolutional layer is the core block of CNN, which contains a set of filters to convolve the input data followed by a nonlinear transformation to extract the spatial features. In the deep learning implementation, there are several key hyper-parameters that should be set in the convolutional layer, like the number of filters, the size of each filter, and so on. The pooling layer generally follows the convolutional layer. The pooling layer aims to reduce the spatial size of the features progressively. In this way, it can help to decrease the number of parameters (e.g., weights and basis) and the computing burden. There are three kinds of pooling operation: max, min, average. Take the max pooling as an example. The pooling operation outputs the maximum value of the pooling area as a result. The hyper-parameters in the pooling layer include the pooling operation, the size of the pooling area, the strides, etc. In the fully connected layer, as in the basic neural network, the nodes have full connections to all activations in the previous layer.

CNN is the most popular deep learning model in brain signal research, which can be used to exploit the latent spatial dependencies among input brain signals like fMRI, spontaneous EEG, and so on.

3.2 Representative Deep Learning Models

The essential blocks of representative deep learning models are AEs and RBMs.[4] DBNs are composed of AE or RBM. The representative models including AE, RBM,[5] and DBN are unsupervised learning methods. Thus, they can learn the representative features from only the input observations x without the ground truth y. In short, representative models receive the

---

[4] AE and RBM are generally regarded as kinds of deep learning although they only have three and two layers, respectively.

[5] We regard AE and RBMas representative methods as most researchers in brain signals adopt them for feature representation.

input data and output a dense representation of the data. There are various definitions in different studies for several models (such as DBN, Deep-RBM, and Deep-AE), the most understandable definitions are chosen and will be presented in detail in this section.

3.2.1　*Autoencoder*

As shown in Figure 3.5(a), an AE is a neural network that has three layers: the input layer, the hidden layer, and the output layer. It differs from the

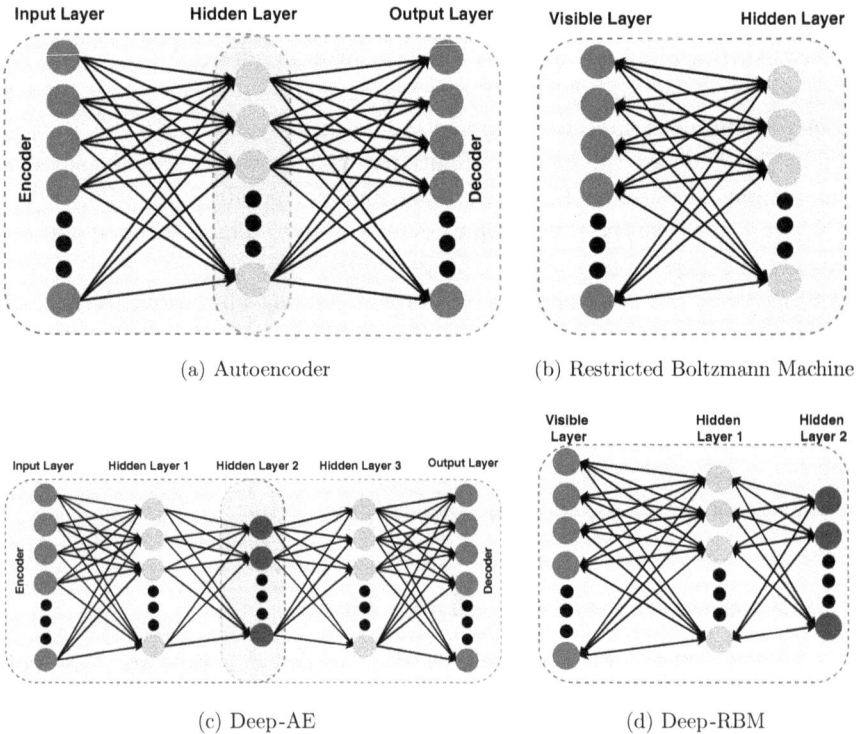

(a) Autoencoder　　　　　　　　(b) Restricted Boltzmann Machine

(c) Deep-AE　　　　　　　　(d) Deep-RBM

Fig. 3.5: Illustration of several standard representative deep learning models. (a) A basic AE contains three layers where the input layer and the output layer are supposed to have the same values. (b) In the RBM, the encoder and the decoder share the same transformation weights. The input layer and the output layer are merged into the visible layer. (c) The stacked AE has more than one hidden layer. Generally, the number of hidden layers is odd, and the middle layer is the learned representative features. (d) The deep-RBM has one visible layer and multiple hidden layers, the last layer is the encoded representation.

standard neural network, in that the AE is trained to reconstruct its inputs, which forces the hidden layer to learn good representations of the inputs.

The structure of AE contains two blocks. The first block is called the encoder, which embeds the observation to a latent representation (also called "code" in some literatures),

$$x^h = \sigma(\mathcal{T}(x)) \tag{3.19}$$

where x^h represents the hidden layer. The second block is called the decoder, which decodes the representation into the original space,

$$y' = \sigma(\mathcal{T}(x^h)) \tag{3.20}$$

where y' represents the output.

AE forces y' to be equal to the input x and calculates the error based on the distance between them. Thus, AE can compute the loss function only by x without the ground truth y

$$error = \|y' - x\|_2 \tag{3.21}$$

Compared to Equation (3.6), this equation does not involve the variable y because it takes the input x as the ground truth. This is the reason why AE is able to perform unsupervised learning.

Naturally, one variant of AE is Deep-AE (D-AE) which has more than one hidden layer. We present the structure of D-AE with three hidden layers in Figure 3.5(c). From the figure, it is observed that there is one more hidden layer in both the encoder and the decoder. The symmetrical structure ensures the smoothness of encoding and decoding procedures. Thus, D-AE generally has an odd number of hidden layers (e.g., $2n + 1$), where the first n layers belong to the encoder, the $(n + 1)$-th layer works as the code that belongs to both encoder and decoder, and the last n layers belong to the decoder. The data flow of D-AE (Figure 3.5(c)) can be represented as

$$x^{h1} = \sigma(\mathcal{T}(x)) \tag{3.22}$$

$$x^{h2} = \sigma(\mathcal{T}(x^{h2})) \tag{3.23}$$

where x^{h2} denotes the median hidden layer (the code). Then, decoding the hidden layer, we can get

$$x^{h3} = \sigma(\mathcal{T}(x^{h2})) \tag{3.24}$$

$$y' = \sigma(\mathcal{T}(x^{h3})) \tag{3.25}$$

It is almost the same as AE except that D-AE has more hidden layers. Apart from D-AE, AE has many other variants like denoising AE, sparse AE, contractive AE, and so on. Here only the D-AE is introduced because it is easily confused with the AE-based DBN. The key difference between them will be provided in Section 3.2.3.

The core idea of AE and its variants is simply condensing the input data x into a code x^h (generally the code layer has lower dimension) and then reconstructing the data based on the code. If the reconstructed y' can approximate the input data x, it can be demonstrated that the condensed code x^h carries enough information about x; thus, x^h can be considered a representation of the input data for future operation (e.g., classification).

3.2.2 *Restricted Boltzmann Machine*

RBM is a stochastic ANN that can learn a probability distribution over its set of inputs. It contains two layers including one visible layer (input layer) and one hidden layer, as shown in Figure 3.5(b). From the figure, we can see that the connection lines between the two layers are bidirectional. RBM is a variant of Boltzmann machine with stronger restriction of being without intra-layer connections.[6] Similar to AE, the procedure of RBM also includes two steps. The first step condenses the input data from the original space to the hidden layer in a latent space. After that, the hidden layer is used to reconstruct the input data in an identical way. Compared to AE, RBM has a stronger constraint which is that the encoder weights and the decoder weights should be equal. We have

$$x^h = \sigma(\mathcal{T}(x)) \qquad (3.26)$$

$$x' = \sigma(\mathcal{T}(x^h)) \qquad (3.27)$$

In the above two equations, the weights of $\mathcal{T}(\cdot)$ are the same. Then, the error for training can be calculated by

$$error = \|x' - x\|_2 \qquad (3.28)$$

It can be observed from Figure 3.5(d) that the Deep-RBM (D-RBM) is an RBM with multiple hidden layers. The input data from the visible layer

---

[6]In a general Boltzmann machine, the nodes in the same hidden layer will connect.

firstly flow to the first hidden layer and then the second hidden layer. Then, the code will flow backward into the visible layer for reconstruction.

3.2.3   *Deep Belief Networks*

A DBN is a stack of simple networks, such as AEs or RBMs (Glauner, 2015). Thus, DBN is divided into DBN-AE (also called stacked AE), which is composed of AE, and DBN-RBM (also called stacked RBM), which is composed of RBM.

As shown in Figure 3.6(a), the DBN-AE contains two AE structures while the hidden layer of the first AE works as the input layer of the second AE. This diagram has two stages. In the first stage, the input data feed into the first AE which follows the rules introduced in Section 3.2.1. The reconstruction error is calculated and back propagated to adjust the corresponding weights and basis. This iteration continues until the AE converges. We get the mapping: $x^1 \rightarrow x^{h1}$.

Then, in the second stage, the learned representative code in the hidden layer x^{h1} will be used as the input layer of the second AE, which is

$$x^2 = x^{h1} \tag{3.29}$$

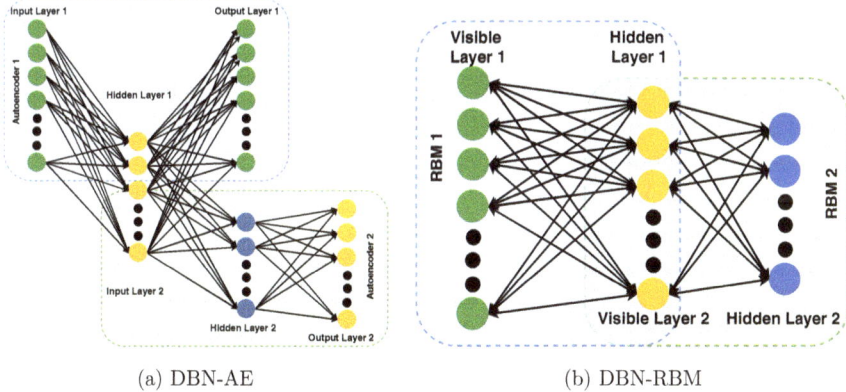

(a) DBN-AE   (b) DBN-RBM

Fig. 3.6: Illustration of DBNs. (a) DBN composed of AEs. DBN-AE contains multiple AE components (in this case, two AE), with the hidden layer of the previous AE working as the input layer of the next AE. The hidden layer of the last AE is the learned representation. (b) DBN composed of RBM. The hidden layer of the first RBM working as the visible layer of the second RBM. The last hidden layer is the encoded representation. While DBN-RBM and D-RBM (Figure 3.5(d)) have similar architecture, the former is trained greedily while the latter is trained jointly.

and then, after the second AE converges, we have

$$x^2 \to x^{h2} \tag{3.30}$$

where x^{h2} denotes the hidden layer of the second AE, meanwhile, it is the final outcome of the DBN-AE.

The core idea of AE is that of learning a representative code with lower dimensionality but containing most information of the input data. The idea behind DBN-AE is to learn a more representative and purer code. Similarly, the DBN-RBM is composed of several single RBM structures. Figure 3.6(b) shows a DBN with two RBMs where the hidden layer of the first RBM is used as the visible layer of the second RBM.

Compare the DBN-RBM (Figure 3.6(b)) and D-RBM (Figure 3.5(d)). They almost have the same architecture. Moreover, DBN-AE (Figure 3.6(a)) and D-AE (Figure 3.5(c)) have similar architecture. The most important difference between the DBN and the deep AE/RBM is that the former is trained greedily, whereas the latter is trained jointly. In particular, for the DBN, the first AE/RBM is trained first, after it converges, the second AE/RBM is trained (Hinton and Salakhutdinov, 2006). For the deep AE/RBM, joint training means that the whole structure is trained together, no matter how many layers it has.

3.3 Generative Deep Learning Models

Generative deep learning models are mainly used to generate training samples or data augmentation. In other words, generative deep learning models play a supporting role in the brain signal field to enhance the training data quality and quantity. After the data augmentation, the discriminative models will be employed for the classification. This procedure is created to improve the robustness and effectiveness of the trained deep learning networks, especially when the training data is limited. In short, the generative models receive the input data and output a batch of similar data. In this section, two typical generative deep learning models, VAE and GAN, will be introduced.

3.3.1 *Variational Autoencoder*

VAE, proposed in 2013 (Kingma and Welling, 2013), is an important variant of AE, and one of the most powerful generative algorithms. The standard AE and its other variants can be used for representation but fail in generation for the reason that the learned code (or representation) may not be continuous. Therefore, a random sample similar to the input sample cannot be generated. In other words, the standard AE does not allow interpolation. Thus, the input sample can be replicated but a similar one cannot be generated. VAE has one fundamentally unique property that separates it from other AEs, and it is this property that makes VAE so useful for generative modeling: the latent spaces are designed to be continuous which allows easy random sampling and interpolation. Next, how VAE works will be explained.

Similar to the standard AE, VAE can be divided into an encoder and decoder where the former embeds the input data to a latent space and the latter transfers the data from the latent space to the original space. However, the learned representation in the latent space is forced to approximate a prior distribution $p(\bar{z})$ which is generally set as standard Gaussian distribution. Based on the re-parameterization trick (Kingma and Welling, 2013), the first hidden layer of VAE is designed to have two parts where one denotes the expectation $\boldsymbol{\mu}$ and another denotes the standard deviation $\boldsymbol{\sigma}$, thus we have

$$\boldsymbol{\mu} = \sigma(\mathcal{T}(\boldsymbol{x})) \tag{3.31}$$

$$\boldsymbol{\sigma} = \sigma(\mathcal{T}(\boldsymbol{x})) \tag{3.32}$$

Then, the latent code in the hidden layer is not directly calculated but sampled from a Gaussian distribution $\mathcal{N}(\boldsymbol{\mu}, \boldsymbol{\sigma}^2)$. The statistic code

$$\boldsymbol{z} = \boldsymbol{\mu} + \boldsymbol{\sigma} * \varepsilon \tag{3.33}$$

where $\varepsilon \sim \mathcal{N}(\boldsymbol{0}, \boldsymbol{I})$. The representation $\boldsymbol{z}$ is forced to a prior distribution, and the distance error_{KL} is measured by Kullback–Leibler divergence,

$$\text{error}_{KL} = D_{KL}(z, \boldsymbol{p}(\bar{\boldsymbol{z}})) \tag{3.34}$$

where $\boldsymbol{p}(\bar{\boldsymbol{z}})$ denotes the prior distribution. In the decoder, $\boldsymbol{z}$ is decoded into the output $\boldsymbol{y}'$,

$$\boldsymbol{y}' = \sigma(\mathcal{T}(\boldsymbol{z})) \tag{3.35}$$

and the reconstruction error is

$$\text{error}_{\text{recon}} = \|y' - x\|_2 \tag{3.36}$$

The overall error for VAE is combined by the DL divergence and the reconstruction error,

$$\text{error} = \text{error}_{KL} + \text{error}_{\text{recon}} \tag{3.37}$$

The key point of VAE is that all the latent representations z are forced to obey the normal distribution. Thus, a representation z' can be randomly sampled from the prior distribution and then a sample based on z' can be reconstructed. This is why VAE is so powerful in generation.

3.3.2   Generative Adversarial Networks

GAN (Goodfellow *et al.*, 2014) was proposed in 2014 and has achieved great success in a wide range of research areas (e.g., computer vision and natural language processing). GAN is composed of two simultaneously trained neural networks with a generator and a discriminator. The generator captures the distribution of the input data, and the discriminator is used to estimate the probability that a sample came from the training data. The generator aims to generate fake samples, whereas the discriminator aims to distinguish whether the sample is genuine. The functions of the generator and the discriminator are opposed; that's why GAN is called "adversarial." After the convergence of both the generator and the discriminator, the discriminator ought to be unable to recognize the generated samples. Thus, the pretrained generator can be used to create a batch of samples and use them for further operations such as as classification.

Figure 3.7(b) shows the procedure of a standard GAN. The generator receives a noise signal s, which is randomly sampled from a multimodal Gaussian distribution and outputs the fake brain signals x_F. The discriminator receives the real brain signals x_R and the generated fake sample x_F, and then it predicts whether the received sample is real or fake. The internal architectures of the generator and discriminator are designed depending on the data types and scenarios. For instance, we can build the GAN by convolutional layers on fMRI images because CNN has an excellent ability to extract spatial features. The discriminator and the generator are trained jointly. After the convergence, numerous brain signals x_G can be created by the generator. Thus, the training set is enlarged from x_R to $\{x_R, x_G\}$ to train a more effective and robust classifier.

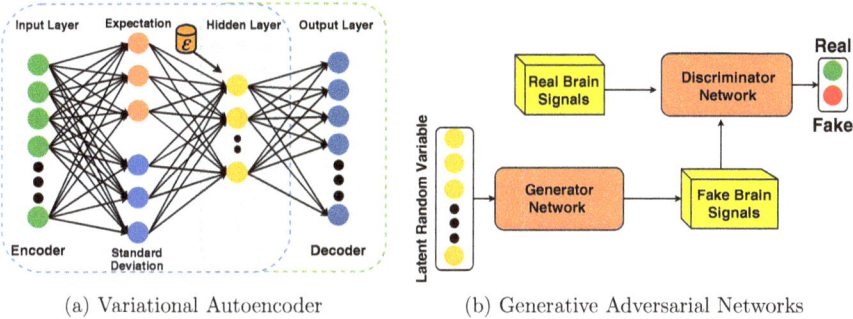

(a) Variational Autoencoder   (b) Generative Adversarial Networks

Fig. 3.7: Illustration of generative deep learning models. (a) VAE contains two hidden layers. The first hidden layer is composed of two components: the expectation and the standard deviation, which are learned separately from the input layer. The second hidden layer represents the encoded information, whereas ϵ denotes the standard normal distribution. (b) GAN mainly contains two crucial components: the generator and the discriminator network. The former receives a latent random variable to generate a fake brain signal, whereas the latter receives both the real and the generated brain signals and attempts to determine if it is generated or not. In brain research, GAN reconstructs or augments data instead of classification.

3.4 Hybrid Models

"Hybrid deep learning models" refers to models that are composed of at least two basic deep learning models, where the basic model is a discriminative, representative, or generative deep learning model. Hybrid models comprise two subcategories based on their targets: classification-aimed (CA) hybrid models and nonclassification-aimed (NCA) hybrid models.

Most deep-learning-related studies in brain research focus on the first category. Based on the existing literature, the representative and generative models are employed to enhance the discriminative models. The representative models can provide more informative and low-dimensional features for discrimination, whereas the generative models can help to augment the training data quality and quantity which supply more information for the classification. The CA hybrid models can be further subdivided into[7]: (1) several discriminative models combined to extract more distinctive and robust features (e.g., CNN+RNN); (2) a representative model followed by a discriminative model (e.g., DBN+MLP); (3) a generative model

---

[7]The representative model followed by a nondeep learning classifier is regarded as a representative deep learning model.

combined with a representative model followed by a discriminative model; and (4) a generative model combined with a representative model followed by a nondeep learning classifier.

A few NCA hybrid models aim for brain signal reconstruction. For example, St-Yves and Naselaris (2018) attempted to recover a complete image based on user's fMRI images. It employed the architecture of GAN to learn a generative model of images that is conditioned on the measurements of brain activity, which can increase the size of training data and can also make the model robust to noise.

PART 2

Deep Learning-Based BCI and Its Applications

Chapter 4

Deep Learning-Based BCI

In this chapter, we will systematically summarize the existing state-of-the-art deep learning techniques in BCI research. Some studies combining deep learning and traditional machine learning methods will also be covered.

4.1   Intracortical and ECoG

In this section, we briefly present the big picture of deep learning-based invasive brain signal analysis. Currently, intracortical brain signals are mainly investigated by medical and biological researchers (Anumanchipalli *et al.*, 2019; Bashivan *et al.*, 2019). Antoniades *et al.* (2016) employed CNN with two convolutional layers to extract features from epileptic intracortical data for detecting interictal epileptic discharge (IED). They sliced the input data into 80 ms segments with 40 ms overlapping windows and achieved an accuracy of about 87.51%. Their later work (Antoniades *et al.*, 2018) used a deep neural architecture to map the scalp signals to pseudo-intracranial brain signals.

Most ECoG-related studies focused on medical healthcare, especially epileptic seizure diagnosis. For example, Hosseini *et al.* (2017a) worked on seizure diagnosis and localization using bandpass filtered (0.1–30 Hz) EEG and ECoG signals. The authors manually extracted features using principal component analysis (PCA), independent component analysis (ICA), and differential search algorithm (DSA), and then fed the features to two different deep learning structures (CNN and DBN-AE) and reported that CNN achieved a higher accuracy (Hosseini *et al.*, 2017a).

Kiral-Kornek *et al.* (2018) proposed an MLP algorithm for an epileptic seizure prediction system operable as a wearable device with an ultra-low power requirement and achieved a sensitivity of 69%. Aside from the studies focused on seizure diagnosis, Xie *et al.* (2018) focused on the finger trajectory from ECoG signals and developed a hybrid deep neural network based on a convolutional layer and LSTM cells. The contribution made by this work lies on its use of CNN not only for spatial convolution but also for temporal convolution as well. The convolution operation produced fixed-length vector representations that were sent to the LSTM cell for trajectory tracking. Each ECoG segment lasted for 1 s with 40 ms overlap. Thus, the model received a stream of ECoG from which it was able to form a complete finger trajectory.

4.2   EEG Potentials

More than half of the recent publications take EEG signals as input because the acquisition system is noninvasive, high-portable, and low-price. In this section, we summarize two aspects of EEG signals: EEG oscillatory and evoked potentials where the former are spontaneous and the latter require outside stimuli.

4.2.1   *Spontaneous EEG Potentials*

The deep learning models for spontaneous EEG according to their application scenarios are presented as follows.

4.2.1.1   *Sleep EEG*

Sleep EEG is mainly used for identifying the sleep stage and diagnosing sleep disorders or cultivating healthy habits (Chambon *et al.*, 2018; Zhang *et al.*, 2016a). According to Rechtschaffen and Kales (R&K) rules, the sleep stages can be categorized as non-rapid eye movement (REM) 1, non-REM 2, non-REM 3, non-REM 4, and REM. The American Academy of Sleep Medicine (AASM) prefers a five-stage model: wakefulness, non-REM 1, non-REM 2, slow wave sleep (SWS), and REM. They combine non-REM 3 and non-REM 4 in SWS since no clear distinction can be made between them (Zhang *et al.*, 2016a). Generally, in sleep stage analysis, the

EEG signals are preprocessed by a filter with a passband that varies by paper, but which is always notched at 50 Hz. The EEG signals are usually segmented into 30 s windows.

(i) *Discriminative models.* CNN are frequently used for sleep stage classification on single-channel EEG (Sors *et al.*, 2018; Tsinalis *et al.*, 2016). For example, Vilamala *et al.* (2017) manually extracted the time–frequency features and achieved a classification accuracy of 86%. The EEG signal stream, collected from $Fpz - Cz$ and $Pz - Oz$ channels, was sliced into 30 s segments. The employed CNN achieved an accuracy of 86% in five-class classification. Shahin *et al.* (2017) manually extracted 57 features from the frequency domain and fed them into an MLP for classification, which obtained an accuracy of 90% in insomnia detection. Fernández-Varela *et al.* (2018) adopted CNN to analyze the physiological signals including EEG, EOG, and EMG. The model was evaluated over the Sleep Heart Health Scoring data set and achieved a precision of 91%, a recall of 90%, and an F-1 score of 90%. Others used RNN (Biswal *et al.*, 2017) and LSTM (Tsiouris *et al.*, 2018) based on various features from the frequency domain, correlation, and graph theoretical features.

(ii) *Representative models.* Tan *et al.* (2015a) adopted a DBN-RBM algorithm to detect sleep spindle based on PSD features extracted from sleep EEG signals and achieved an F1 of 92.78% on a local data set. Zhang *et al.* (2016a) further combined DBN-RBM with three RBMs for sleeping feature extraction.

(iii) *Hybrid models.* Manzano *et al.* (2017) proposed a multiview algorithm in order to predict sleep stage by combining CNN and MLP. The CNN was employed to receive the raw time-domain EEG oscillations, whereas the MLP received the spectrum signals of 0.5–32 Hz obtained by the Short-Time Fourier Transform (STFT). Fraiwan and Lweesy (2017) combined DBN with MLP for neonatal sleep state identification. Supratak *et al.* (2017) proposed a model by combing a multiview CNN and LSTM for automatic sleep stage scoring, in which the former was adopted to discover time-invariant dependencies while the latter (a bidirectional LSTM) was adopted the identify the temporal features of the sleep spindles. Dong *et al.* (2018) proposed a hybrid deep learning model aimed at temporal sleep stage classification which employed MLP for detecting hierarchical features with LSTM for sequential information learning.

4.2.1.2   *Motor-Imagery EEG*

Deep learning models have been shown to be superior in the classification of EEG and real-motor EEG (Hartmann *et al.*, 2018; Nurse *et al.*, 2016).

(i) *Discriminative models.* Such models mostly use CNN to classify MI EEG (Zhang *et al.*, 2019b). Some are based on manually extracted features (Jingwei *et al.*, 2015; Yang *et al.*, 2015). For instance, Lee and Choi (2018) and Zhang *et al.* (2017b) employed CNN and 2D CNN, respectively, for classification; Zhang *et al.* (2017b) learned affective information from EEG signals to build a modified LSTM to control smart home appliances. Others also used CNN for feature engineering (Tang *et al.*, 2017). For example, Wang *et al.* (2017) first used CNN to capture latent connections from MI EEG signals and then applied weak classifiers to choose important features for the final classification; Hartmann *et al.* (2018) investigated how CNN represented spectral features through the sequence of the MI EEG samples. MLP has also been applied for MI EEG recognition (Sturm *et al.*, 2016), and showed higher sensitivity to EEG phase features at earlier stages and higher sensitivity to EEG amplitude features at later stages.

(ii) *Representative models.* DBN is widely used as a basis for MI EEG classification for its high representative ability (Kumar *et al.*, 2016; Lu *et al.*, 2017). For example, Ren and Wu (2014) applied a convolutional DBN based on RBM components, showing better feature representation than hand-crafted features; Li and Cichocki (2014) processed EEG signals with discrete wavelet transformation and then applied a DBN-AE based on denoising AE. Other models include the combination of an AE model (for feature extraction) with a KNN classifier (Redkar, 2015) the combination of a Genetic Algorithm (for hyperparameter tuning) with an MLP (for classification) (Nurse *et al.*, 2015), and the combination AE and XGBoost for multiperson scenarios (Zhang *et al.*, 2017c). Nurse *et al.* (2015) proposed a model combining an MLP with a Genetic Algorithm (GA) in which the GA was used for optimal hyperparameter selection (e.g., the number of hidden layers in MLP) and the MLP worked as the classifier. Zhang *et al.* (2017c) combined an AE with an XGBoost classifier to recognize the EEG signals in a multiperson scenario.

(iii) *Hybrid models.* Several studies proposed hybrid models for the recognition of MI EEG (Dai *et al.*, 2019). For example, Tabar and Halici (2016) extracted high-level representations from the time–frequency domain and location information of EEG signals using CNN and then applied a

DBN-AE with seven AEs as the classifier; Tan *et al.* (2017) used a denoising AE for dimensional reduction with a multiview CNN combined with RNN to discover latent temporal and spatial information, and achieved an average accuracy of 72.22% on a public data set.

(iv) *Event-related desynchronization/synchronization (ERD/ERS)*. To make our review of the literature more comprehensive, a brief introduction to ERD/ERS is provided here. ERD/ERS refers to the phenomena that the magnitude and frequency distribution of the EEG signal power changes during a specific brain state (Huang *et al.*, 2012). In particular, ERD denotes the power decrease of ongoing EEG signals whereas ERS represents the power increase of EEG signals. This characteristic of ERD/ERS of brain signals can be used to detect the event which caused the EEG fluctuation. For example, Pfurtscheller and Da Silva (1999) present the ERD/ERS phenomena in the motor cortex recorded during a motor-imagery task.

ERD/ERS are mainly observed in sensory, cognitive and motor procedures, which are not widely used in brain research due to the drawbacks involved like inconsistent performance across subjects (Huang *et al.*, 2012). In most situations, ERD/ERS is regarded as a specific feature of the EEG powers for further analysis (Dai *et al.*, 2019; Tabar and Halici, 2016). The task causes an ERD in the mu band (8–13 Hz) of EEG and an ERS in the beta band (13–30 Hz). In particular, the ERD/ERS were calculated as power changes relative to a baseline: $ERD/ERS = (P_e - P_b)/P_b$, where P_e denotes the signals' power over a one-second segment while the event occurring and P_b denotes the signal power in a one-second segment during baseline which is before the event (Chiarelli *et al.*, 2018). Generally, the baseline refers to the rest state. For example, Sakhavi *et al.* calculated the ERD/ERS map and analyzed the different patterns produced during various tasks. The analysis demonstrated that the dynamic of energy should be considered because the static energy does not contain enough information (Sakhavi *et al.*, 2015).

4.2.1.3 *Emotional EEG*

An individual environment state can be evaluated for three aspects: valence, arousal, and dominance. These three aspects combined produce human emotions such as fear, sadness, and anger, and can be read in emotional EEG signals.

(i) *Discriminative models.* MLP are traditionally used (Frydenlund and Rudzicz, 2015; Yepes *et al.*, 2017) while CNN and RNN are increasingly popular in EEG-based emotion prediction (Li *et al.*, 2016a; Liu *et al.*, 2017b). Typical CNN-based work in this category includes hierarchical CNN modeling (Li *et al.*, 2016a, 2017a) and augmenting the training set for CNN (Wang *et al.*, 2018). Li *et al.* (2016a) were the first to propose capturing the spatial dependencies among EEG channels by converting multichannel EEG signals into a 2D matrix. Talathi (2017) used a discriminative deep learning model composed of GRU cells. Zhang *et al.* (2018b) proposed a spatial temporal RNN employing a multidirectional RNN layer to discover long-range contextual cues and a bidirectional RNN layer to capture the temporal features of the sequences produced by the previous spatial RNN.

(ii) *Representative models.* DBN, especially DBN-RBM, is widely used in the unsupervised representation learning for emotion recognition (Gao *et al.*, 2015b; Li *et al.*, 2013, 2015b). For instance, Xu and Plataniotis (2016a,b) proposed a DBN-RBM algorithm with three RBMs and an RBM-AE to predict affective state; Zhao and He (2014) and Zheng *et al.* (2014) combined a DBN-RBM with a support vector machine (SVM) and a hidden Markov model (HMM), respectively, to address the same problem; Zheng and Lu (2015) and Zheng *et al.* (2015) introduced a D-RBM with five hidden RBM layers to search the important frequency patterns and informative channels in affection recognition; Jia *et al.* (2014) first eliminated channels with high errors, then used D-RBM for affective state recognition based on representative features of the residual channels.

Emotions may be affected by many subjective and environmental factors (e.g., gender and fatigue). Liu *et al.* (2016) investigated the emotional signals of men and women for patterns of discrepancies using a novel model called bimodal deep autoencoder (BDAE) which received both EEG and eye movement features and shared the information in a fusion layer which connected with an SVM classifier. The females' results showed higher EEG signal diversity for the fearful emotion, whereas the males varied more for sadness. Furthermore, for the women, the intersubject differences in fear were more significant than for other emotions (Liu *et al.*, 2016). To overcome the mismatched distribution among the samples collected from different subjects or different experimental sessions, Chai *et al.* (2016) proposed an unsupervised domain adaptation technology called the subspace alignment autoencoder (SAAE), which combines an AE and a subspace alignment solution. The proposed approach obtained a mean accuracy of 77.88% in a person-independent scenario.

(iii) *Hybrid models.* One commonly used hybrid model is a combination of RNN and MLP. For example, Alhagry *et al.* (2017) employed an LSTM architecture for feature extraction from emotional EEG signals and the learned features were forwarded to an MLP for classification. Furthermore, Yin *et al.* (2017) proposed a multiview ensemble classifier to recognize individual emotions using multimodal physiological signals. The ensemble classifier contained several D-AEs, three hidden layers and a fusion structure. Each D-AE received one physiological signal (e.g., EEG, EOG, EMG) and sent its outputs to a fusion structure composed of another D-AE. Finally, an MLP classifier was employed to classify the mixed features. Kawde and Verma (2017) implemented an affect recognition system by combining a DBN-RBM for effective feature extraction and an MLP for classification.

4.2.1.4 *Mental Disease EEG*

A large number of researchers used EEG signals to diagnose neurological disorders, especially epileptic seizure (Yuan *et al.*, 2018).

(i) *Discriminative models.* The CNN model is widely used in the automatic detection of epileptic seizure (Acharya *et al.*, 2018a; Schirrmeister *et al.*, 2017; Ullah *et al.*, 2018; Wang *et al.*, 2016b). For example, Johansen *et al.* (2016) adopted CNN to work on the high-pass filtered (1 Hz) EEG signals of epileptic spike and achieved an AUC of 94.7%. Acharya *et al.* (2018b) employed a CNN model with 13 layers to screen for depression, which was evaluated on a local data set with 30 subjects and achieved accuracy rates of 93.5% and 96.0% based on the left- and right-hemisphere EEG signals, respectively. Using a CNN structure, Morabito *et al.* (2016) tried to extract suitable features of multichannel EEG signals to distinguish patients with Alzheimer's disease from patients with mild cognitive impairment and age-matched healthy controls. The EEG signals were filtered in bandpass (0.1–30 Hz) and achieved an accuracy of around 82% for the three-class classification. Rapid eye movement behavior disorder (RBD) predicts and may cause many psychiatric disorders/diseases such as Parkinson's disease (PD). Ruffini *et al.* (2016) described an echo state network (ESN) model, a particular class of RNN, to distinguish those with RBD from healthy individuals. In some research, the discriminative model is employed only for feature extraction. Ansari *et al.* (2018) used CNN to extract latent features which were then fed into a random forest classifier for seizure detection in neonatal babies. Chu *et al.* (2017) combined CNN and a traditional classifier for schizophrenia recognition.

(ii) *Representative models.* For disease detection, one widely used method is the adoption of a representative model (e.g., DBN) followed by a softmax layer for diagnosis (Turner *et al.*, 2014). Page *et al.* (2014) adopted DBN-AE to extract informative features that were fed into a traditional logistic regression classifier for seizure detection. Al-kaysi *et al.* (2015) proposed a multiview DBN-RBM structure to analyze the EEG signals of patients with depression. Their approach contained multiple input pathways, each composed of two RBMs, and each of which corresponded to a single EEG channel. All of the input pathways merged into a shared structure composed of a final RBM. Some models used dimensionality reduction methods such as PCA for preprocessing of the EEG signal (Hosseini *et al.*, 2017b), whereas others chose to directly fed the raw signals to the representative model (Lin *et al.*, 2016). Lin *et al.* (2016) proposed a sparse D-AE with three hidden layers to extract the representative features from epileptic EEG signals while Hosseini *et al.* (2017b) adopted a similar sparse D-AE with only two hidden layers.

(iii) *Hybrid models.* One popular hybrid method uses a combination of RNN and CNN. Shah *et al.* (2017) tested the performance of CNN-LSTM in detecting seizures after channel selection; sensitivity ranged from 33% to 37%, and false alarms ranged from 38% to 50%. Golmohammadi *et al.* (2017c) proposed a hybrid architecture for automatic interpretation of EEG through integration of both the temporal and spatial information. 2D and 1D CNNs captured the spatial features while LSTM networks captured the temporal features. The authors claimed a sensitivity of 30.83% and a specificity of 96.86% on the well-known TUH EEG seizure corpus. In the detection of early-stage Creutzfeldt-Jakob disease (SJD), Morabito *et al.* (2017) combined D-AE and MLP together. The EEG signals of patients with SJD were first filtered by bandpass (0.5–70 Hz) and then fed into a D-AE with two hidden layers for feature representation. The MLP classifier obtained an accuracy of 81–83% on a local data set. Wen and Zhang (2018) applied a convolutional autoencoder, in which the fully connected layers in a standard AE are replaced with convolutional and de-convolutional layers, to extract the seizure features in an unsupervised manner (Wen and Zhang, 2018).

4.2.1.5 *Data Augmentation*

It has been proposed that the generative models such as GAN could be used for data augmentation in brain signal classification (Abdelfattah *et al.*, 2018). Palazzo *et al.* (2017) first demonstrated that the information contained in the brain waves allows identification of the visual object from which a more robust and distinguishable representation of EEG data can be extracted using RNN. The GAN paradigm was employed to train an image generator conditioned by these learned EEG representations and was able to convert the EEG signals into images (Palazzo *et al.*, 2017). Kavasidis *et al.* (2017), aiming at converting EEG signals into images, collected the signals while subjects observed images on a screen. An LSTM layer was employed to extract the latent features from the EEG signals, and the extracted features were input into a GAN structure. The GAN generator and the discriminator were both composed of convolutional layers. The generator was intended to generate an image based on the input EEG signals after the pre-training. Abdelfattah *et al.* (2018) adopted a GAN on seizure data augmentation. The generator and discriminator were both composed of fully connected layers. The authors demonstrated that GAN outperforms both AE and VAE. After the augmentation, classification accuracy increased dramatically from 48% to 82%.

4.2.1.6 *Other Models*

A wide range of exciting topics have been explored by researchers in this field. The first of these is how EEG is affected by audio/visual stimuli. In this phenomenon, the stimuli are continuous, so that they exist in a steady state instead of as flickering potentials. Stober *et al.* (2014a,b) claimed that the rhythm-evoked EEG signals are informative enough to distinguish the rhythm stimuli. The authors conducted an experiment where 13 participants were stimulated by 23 rhythmic stimuli, including 12 East African and 12 Western stimuli. For the 24-category classification, the proposed CNN achieved a mean accuracy of 24.4%. After that, the authors exploited convolutional AE for representation learning and CNN for recognition and achieved an accuracy of 27% for 12-class classification (Stober *et al.*, 2015). Sternin *et al.* (2015) adopted CNN to capture discriminative features from the EEG oscillations to distinguish whether the subject was listening or imaging music. Similarly, Sarkar *et al.* (2016) designed two deep learning models to recognize the EEG signals aroused by audio or visual stimuli. For this binary classification task, the proposed CNN and DBN-RBM with three RBMs achieved the accuracy of 91.63% and 91.75%, respectively. Furthermore, the spontaneous EEG could be used to distinguish the user's mental state (logical versus emotional) (Bashivan *et al.*, 2016a).

Other researchers have focused on the impact of cognitive load (Shang *et al.*, 2017) and physical load on EEG (Gordienko *et al.*, 2017). Bashivan *et al.* (2015) first extracted informative features through entropy-based wavelet transformation and band-specific power, which were fed into a DBN-RBM for further refining, then to an MLP for cognitive load level recognition. In other works, the authors (Bashivan *et al.*, 2016b) also claimed to discover the general features which are constant in inter-/intrasubjective scenarios under varying mental loads. Yin and Zhang (2017) collected the EEG signals from different mental workload levels (e.g., high and low) for binary classification. The EEG signals were filtered by a low-pass band filter, transformed to the frequency domain and the power spectral density (PSD) was calculated. The extracted PSD features were fed into a denoising D-AE structure for further refining, which gave an accuracy of 95.48%. Li *et al.* (2016b) worked on the recognition of mental fatigue levels, including alert, slight fatigue, and severe fatigue.

Incidentally, EEG-based driver fatigue detection is another attractive area for study (Chai *et al.*, 2017; Du *et al.*, 2017; Hajinoroozi *et al.*, 2015b). Hung *et al.* (2017) designed a 3D CNN to predict the reaction time for drowsy driving that could prove instrumental in reducing traffic accident. Hajinoroozi *et al.* (2015a) adopted a DBN-RBM to handle the EEG signals which were processed by ICA. They achieved an accuracy of around 85% in binary classification ("drowsy" or "alert"). The strength of this work is that they evaluated the DBN-RBM on three levels: time samples, channel epochs, and windowed samples. The experiments illustrated that the channel epoch level outperformed the other two levels. San *et al.* (2016) combined deep learning models with a traditional classifier to detect driver fatigue. The model contained a DBN-RBM structure followed by an SVM classifier, which achieved a detection accuracy of 73.29%. Almogbel *et al.* (2018) investigated the drivers' mental states under different low workload levels. The authors have claimed their proposed CNN was capable of detecting the driving workload directly from the raw EEG signals.

Research on the detection of eye state has proved exceedingly accurate. Narejo *et al.* (2016) explored the detection of eye state (closed or open) based on EEG signals. They tested a DBN-RBM with three RBMs and a DBN-AE with three AEs and achieved a high accuracy of 98.9%. Reddy and Behera (2016) tried a simpler structure, MLP, and got a slightly lower accuracy of 97.5%.

There are also several overlooked yet promising areas. Baltatzis *et al.* (2017) adopted CNN to detect incidences of school bullying through the EEG signals produced while watching a specific video. They achieved 93.7% and 88.58% for binary and four-class classification, respectively. Khurana *et al.* (2016) proposed deep dictionary learning that outperformed several other deep learning methods. Völker *et al.* (2018) evaluated the use of deep CNN in a flanker task experiment, which achieved an average accuracy of 84.1% within subject and 81.7% on the unseen subjects. Zhang *et al.* (2018c) combined a CNN and a graph neural network to discover the latent information present in the EEG signal.

Miranda-Mioranda-Correa and Patras (2018) proposed a cascaded framework which combined RNN and CNN to predict individuals' affective level and personal factors (Big-Five personality traits, mood, and social context). An experiment conducted by Putten *et al.* (2018) attempted to identify the user's gender based on their EEG signals. They employed a standard CNN algorithm and achieved the binary classification accuracy of 81% on a local data set. The detection of a driver's emergency braking intention could help to reduce the response time. Hernández *et al.* (2018) demonstrated that the driver's EEG signals could communicate braking intention and normal driving state. They combined a CNN algorithm which achieved an accuracy of 71.8% in binary classification. Behncke *et al.* (2018) applied deep learning, a CNN model, in the context of robotic assistive devices. They attempted to use CNN to improve the accuracy of decoding robot errors from EEG while the subject was watching the robot both while grasping an object and performing a pouring task.

Teo *et al.* (2017) tried to combine the brain signal and recommender system, which predicted the user's preference by interpreting EEG signals. Sixteen participants took part in the experiments which collected the EEG signals while the subject was presented with 60 bracelet-like objects as rotating visual stimuli (a 3D object). Then an MLP algorithm was adopted to classify the user's like or dislike of the object. This exploration had a prediction accuracy of 63.99%. A few researchers have tried to establish a common framework which could be used for multiple BCI paradigms. Lawhern *et al.* (2018) introduced EEGNet based on a compact CNN and evaluated its robustness in various brain signal contexts.

4.2.2   *Evoked Potentials*

4.2.2.1   *ERP*

In most situations, ERP signals are analyzed in the context of the P300 phenomena. Meanwhile, almost all the studies on P300 are based on the scenario of ERP. Therefore, in this section, the majority of the P300-related publications are introduced in the subsections of VEP/AEP according to the relevant scenario.

(i) *VEP*. VEP is one of the most popular subcategories of ERP (Haider and Fazel-Rezai, 2017; Spampinato *et al.*, 2017; Yin *et al.*, 2015). Min *et al.* (2017) worked on motion-onset VEP (mVEP) by extracting representative features through deep learning and adopted a genetic algorithm combined with multilevel thresholding structure to compress the raw signals. The compressed signals were sent through a DBN-RBM algorithm to capture the more abstract high-level features. Maddula *et al.* (2017) filtered the P300 signals with visual stimuli by a bandpass filter (2–35 Hz) and then fed the results into a proposed hybrid deep learning model for further analysis. The model included a 2D CNN structure to capture the spatial features followed by an LSTM layer for temporal feature extraction. Liu *et al.* (2017a) combined a DBN-RBM representative model with an SVM classifier for concealed information test and achieved a high accuracy of 97.3% on a local data set. Gao *et al.* (2015a) employed an AE model for feature extraction followed by an SVM classifier. In this experiment, each segment contained 150 points, which were divided into five time steps, where each step had 30 points. This model achieved an accuracy of 88.1% on a local data set. A wide range of P300 speller studies are based on the P300 speller (Shanbhag *et al.*, 2017), which allows the user to write characters. Cecotti (2017) tried to improve the P300's detection accuracy for more precise spelling. A new model was presented based on CNN, which included five low-level CNN classifiers with the different feature set, and the final high-level results are voted by the low-level classifiers. The highest accuracy reached was 95.5% on the data set II from the third BCI competition. Liu *et al.* (2018b) proposed a batch normalized neural network (BN^3) which is a variant of CNN based on the P300 speller consisting of six layers, and for which batch normalization is performed on each batch. Kawasaki *et al.* (2015) employed an MLP model to detect P300 segments from non-P300 segments and achieved an accuracy of 90.8%.

(ii) *AEP*. A few works focused on the recognition of AEP. For example, Carabez *et al.* (2017) proposed and tested 18 CNN structures for classifying single-trial AEP signals. In their experiment, the volunteers were required to wear earphones which produced auditory stimulus patterns based on the oddball paradigm. The experimental analysis demonstrated that each of these CNN frameworks, regardless of the number of their convolutional layers, were effective in extracting both the temporal and spatial features and produced competitive results. The AEP signals were filtered by 0.1–8 Hz and downsampled from 256 to 25 Hz. The experimental results showed that the downsampled data worked better. The SincNet Ravanelli and Bengio (2018) that is usually used in audio processing pays more attention to low and high cutoff frequencies rather than all frequency bands in standard CNN. In particular, it finds some meaningful filters through adding some constraints at the first convolutional layer. Zeng *et al.* (2019) modified SincNet model by including spatial dependencies among EEG channels and improving the filter design at the first layer of SincNet, and applied it to emotional EEG signals analysis. Experimental results show that the improved model has competitive classification accuracy and better algorithm robustness.

(iii) *RSVP*. Among the various VEP diagrams, RSVP has attracted much attention (Gordon *et al.*, 2017). In the analysis of RSVP, a number of discriminative deep learning models (e.g., CNN (Cecotti, 2017; Lin *et al.*, 2017; Solon *et al.*, 2017) and MLP (Mao *et al.*, 2014)) has achieved great success. Frequency filtering is a commonly used preprocessing method for RSVP signals. The passbands used generally range from 0.1 to 50 Hz (Manor and Geva, 2015; Shamwell *et al.*, 2016). Cecotti *et al.* (2014) worked on the classification of ERP signals in the RSVP scenario and proposed a CNN algorithm for the detection of a specific target in RSVP. In the experiment, the images of faces and cars were regarded as target or nontarget, respectively. The image presenting frequency was 2 Hz. In each session, the target probability was 10%. The proposed model offered an AUC of 86.1%. Hajinoroozi *et al.* (2017) adopted a CNN model targeting the intersubject and intertask detection of RSVP. The experimental results showed that CNN worked well in cross-task but failed to get perform satisfactorily in the cross-subject scenario. Mao (2016) compared three different deep neural network algorithms in the prediction as to whether or not the subject had seen the target. The MLP, CNN, and DBN models obtained an AUC of 81.7%, 79.6%, and 81.6%, respectively. The author also applied a CNN model to analyze the RSVP signals for person identification (Mao *et al.*, 2017). Joshi *et al.* (2018) implemented a novel convolutional long short term memory model for single trial P300 classification in order to capture the temporal and spatial information at the same time.

The representative deep learning models are also applied in RSVP. Vařeka and Mautner (2017) tested if deep learning performs well for single trial P300 classification. They conducted an RSVP experiment in which the subjects were asked to distinguish the target from nontargets and distracters. Then a DBN-AE was implemented and compared with some nondeep learning algorithms. The DBN-AE was composed of five AEs where the hidden layer of the last AE had only two nodes capable of classification using the softmax function. The proposed model achieved an accuracy of 69.2%. Manor *et al.* (2016) applied two deep neural networks to deal with the RSVP signals after lowpass filtering (0–51 Hz). Discriminative CNN achieved an accuracy of 85.06%. Meanwhile, the representative convolutional D-AE achieved an accuracy of 80.68%.

4.2.2.2 *SSEP*

Most of the deep learning-based studies in the SSEP area focus on SSVEP. SSVEP refers to brain oscillations evoked by flickering visual stimuli, which are generally produced in the parietal and occipital regions (Waytowich *et al.*, 2018). Attia *et al.* (2018) aimed at finding an intermediate representation of SSVEP. They used a hybrid method of combined CNN and RNN to capture the meaningful features directly from the time domain, which achieved an accuracy of 93.59%. Waytowich *et al.* (2018) applied a compact CNN model to work directly on the raw SSVEP signals without any handcrafted features. The reported cross-subject mean accuracy was approximately 80%. Thomas *et al.* (2017) first filtered the raw SSVEP signals through a bandpass filter (5–48 Hz) and then applied discrete FFT on 512 consecutive points. The processed data were independently classified by a CNN (69.03%) and an LSTM (66.89%) independently.

Pérez-Benítez *et al.* (2018) adopted a representative model, a sparse AE, to extract the distinct features from the SSVEP from multifrequency visual stimuli. The proposed model employed a softmax layer for the final classification and achieved an accuracy of 97.78%. Kulasingham *et al.* (2016) classified SSVEP signals in the context of the guilty knowledge test. The authors applied DBN-RBM and DBN-AE independently and achieved respective accuracies of 86.9% and 86.01%. HACHEM *et al.* (2014) investigated the influence of fatigue on SSVEP through an MLP model during wheelchair navigation. The goal of this study was to seek the key parameters for switching between manual, semi-autonomous, and autonomous wheelchair control. Aznan *et al.* (2018) explored accuracy of SSVEP classification where the signals were collected through dry electrodes. The dry signals were more challenging for the lower SNR than standard EEG signals. This study applied a CNN discriminative model and the highest accuracy achieved was 96% on a local data set. Nguyen and Chung (2018) employed a one-dimension CNN model to detect single-channel SSVEP frequency and achieved a very high performance both online and offline. Ravi *et al.* (2020) manually extracted SSVEP features (magnitude spectrum features and complex spectrum features) and then fed them into a CNN classifier, which achieves competitive results in person-dependent and person-independent scenarios although the used CNN model is rather simple (two convolutional layers followed by a fully connected layer).

4.3 fNIRS

Up to now, only a few of researchers have interested themselves in deep learning–based fNIRS. Naseer *et al.* (2016) analyzed the difference between two mental states (mental arithmetic and rest) based on fNIRS signals. The authors manually extracted six features from the prefrontal cortex fNIRS and used them to compare the performance of six classifiers. The results demonstrated that the MLP classifier, with an accuracy of 96.3% outperformed all of the traditional classifiers, including SVM, KNN, naive Bayes, and so on. Huve *et al.* (2017) classified fNIRS signals collected from the subjects during three mental states (subtractions, word generation, and rest) using an MLP model that achieved an accuracy of 66.48% based on handcrafted features (e.g., the concentration of OxyHb/DeoxyHb). In a subsequent papers, the authors studied the mobile robot control through fNIRS signals and got a binary classification accuracy of 82% (offline) and 66% (online) (Huve *et al.*, 2018). Chiarelli *et al.* (2018) used a combination of fNIRS and EEG for left/right MI EEG classification. Sixteen features extracted from fNIRS signals (eight from OxyHb and eight from DeoxyHb) were fed into an MLP classifier with four hidden layers.

Focusing on another area, Hiroyasu *et al.* (2014) attempted to detect the gender of a subject through analysis of their fNIRS signals. The authors employed a denoising D-AE with three hidden layers to extract distinctive features that were then fed into an MLP classifier for gender detection. The model was evaluated on a local data set and gave an average accuracy of 81%. The authors also pointed out that, compared with PET and fMRI, fNIRS has a higher time resolution and is more affordable (Hiroyasu *et al.*, 2014).

4.4 fMRI

Recently, several deep learning methods have been applied to fMRI analysis, especially in the diagnosis of cognitive impairment (Vieira *et al.*, 2017; Wen *et al.*, 2018).

(i) *Discriminative models.* Among the discriminative models, CNN looks promising for analyzing fMRI (Shreyas and Pankajakshan, 2017). For example, Havaei *et al.* built a segmentation approach for brain tumor based on fMRI with a novel CNN algorithm able to simultaneously capture both the global and local features (Havaei *et al.*, 2017). The convolutional filters were of different size, thus, the small-size and large-sized filters could exploit the local and global features independently. Sarraf *et al.* (2016) and Sarraf and Tofighi (2016) applied deep CNN to recognize Alzheimer's disease based on fMRI and MRI data. Marc Moreno (2017) employed a CNN model to deal with fMRI of brain tumor patients for three-class recognition (normal, edema, or active tumor). The model was evaluated on a BRATS data set and obtained the F1 score of 88%. Hosseini *et al.* (2017c) employed CNN for feature extraction. The extracted features were classified by SVM for the detection of epileptic seizure.

Li *et al.* (2014) had previously proposed a data completion method based on CNN which utilized the information from fMRI data to complete positron emission tomography (PET), then train the classifier based on both fMRI and PET. In this model, the input data of the proposed CNN is the fMRI patch, and the output is a PET patch. Two convolutional layers with 10 filters mapped the fMRI to PET. The authors' experiments illustrated that the classifier trained on the combination of fMRI and PET (92.87%) outperformed the one trained by solo fMRI (91.92%) Moreover, Koyamada *et al.* used a nonlinear MLP to extract common features from different subjects. The model was evaluated on a data set from the Human Connectome Project (HCP) (Koyamada *et al.*, 2015).

(ii) *Representative models.* A wide range of publications have demonstrated the effectiveness of representative models in classifying of fMRI data (Suhaimi *et al.*, 2015). Hu *et al.* (2016) demonstrated that deep learning outperforms other machine learning methods in the diagnosis of neurological disorders such as Alzheimer's disease (AD). First, the fMRI images were converted to a matrix to represent the activity of 90 brain regions. Second, a correlation matrix was obtained by calculating the correlation between each pair of brain regions to represent the functional connectivity between different regions. Finally, a targeted AE sensitive to AD was built to classify the correlation matrix. The proposed approach achieved an accuracy of 87.5%. Plis *et al.* (2014) employed a DBN-RBM with three RBM components to extract the distinctive features from ICA processed fMRI and finally achieved an average F1 measure of above 90% on four public data sets. Suk *et al.* compared the effectiveness of DBN-RBM against DBN-AE on AD detection and the experimental results showed that the former obtained the accuracy of 95.4%, slightly lower than the latter at 97.9% (Suk *et al.*, 2015). Suk *et al.* (2016) applied a D-AE model to extract latent features from the resting-state fMRI data in the diagnosis of mild cognitive impairment (MCI). The latent features were fed into an SVM classifier which achieved an accuracy of 72.58%. Ortiz *et al.* (2016) proposed a multiview DBN-RBM to receive MRI and PET data simultaneously. The learned representations were sent to an ensemble of several simple SVM classifiers which formed a high-level stronger classifier by voting.

(iii) *Generative models.* The reconstruction of natural images (e.g., fMRI) has attracted lots of attention (Shen *et al.*, 2019; Han *et al.*, 2018; Zhang *et al.*, 2018b). Seeliger *et al.* (2018) proposed a deep convolutional GAN for reconstructing visual stimuli from fMRI, using a pretrained generator to create an image similar to the presented stimulus image. The generator required four convolutional layers to convert the input fMRI to a natural image. Han *et al.* (2018) focused on the generation of synthetic multisequence fMRI using GAN. The generated image could be used for data augmentation as well as physician training for improved diagnostic accuracy. In real studies, the authors applied the existing deep convolutional GAN (DCGAN) (Radford *et al.*, 2016) and Wasserstein GAN (WGAN) (Arjovsky *et al.*, 2017) and found that the former works better. Shen *et al.* (2019) presented another image recovery approach which minimized the distance between the real image and the image generated based on real fMRI.

4.5 EOG

In most situations, the EOG signals are regarded as artifacts to be removed from the collected EEG. However, they could also be used as informative signals that deploy on EOG-based BCI system. Although there were some researchers who considered EOG analysis, only a limited number of papers utilized deep learning. For example, Xia *et al.* (2015) attempted to identify the subjects' sleep stage using only EOG signals. They employed a DBN-RBM for feature representation and an HMM for classification. On the other hand, EOG has been widely used to supplement of other signals (e.g., EEG) in studies covering several research topics such as emotion detection (Kawde and Verma, 2017; Liu *et al.*, 2016; Wang *et al.*, 2016b), sleep stage recognition (Fernández-Varela *et al.*, 2018; Supratak *et al.*, 2017), and driving fatigue detection (Du *et al.*, 2017).

4.6   MEG

Garg *et al.* (2017) worked on refining MEG signals by first removing artifacts like eyeblinks and cardiac activity. The MEG signals were decomposed by ICA, then classified by a 1D CNN model. The proposed approach achieved a sensitivity of 85% and a specificity of 97% on a local data set. Hasasneh *et al.* (2018) also focused on artifact detection (cardiac and ocular artifacts). Their proposed approach used CNN to capture temporal features and MLP to extract spatial information. Shu and Fyshe (2013) employed a sparse AE to learn the latent dependencies of MEG signals in the task of single word decoding. The results demonstrated that the proposed approach is advantageous for some subjects, although it did not produce an overall increase in decoding accuracy. Cichy *et al.* (2016) applied a CNN model to recognize visual objects based on MEG and fMRI signals.

4.7   Discussion

Next, we attempt to provide some guidelines for the design of deep learning-based BCI systems derived from our analysis of the state-of-the-art studies. We first determine which are the most suitable deep learning models for each type of brain signal. Next, we summarize the most popular deep learning models in current brain signal research. We hope these guidelines will help our readers to select the most effective and efficient methods when dealing with brain signals. Table 4.1, summarizes the brain signal acquisition methods and the corresponding deep learning models used in state-of-the-art papers. The hybrid model category includes three types: combination RNN/CNN, the combination of representative/discriminative

Table 4.1: A summary of brain signal studies based on deep learning models.

Brain Signals			Deep Learning Models		
			Discriminative Models		
			MLP	RNN	CNN
Non-invasive Signals	EEG	Invasive	Kiral-Kornek et al. (2018)		Antoniades et al. (2016); Hosseini et al. (2017a)
		Spont-aneous EEG — Sleep EEG	Shahin et al. (2017), Biswal et al. (2017)	Tsiouris et al. (2018), Biswal et al. (2017)	Vilamala et al. (2017), Chambon et al. (2018), Tsinalis et al. (2016); Sors et al. (2018), Fernández-Varela et al. (2018); Biswal et al. (2017)
		MI EEG	Chiarelli et al. (2018), Sturm et al. (2016)	Zhang et al. (2018g), Zhang et al. (2019b, 2017b)	Lee and Choi (2018), Uktveris and Jusas (2017), Nurse et al. (2016), Jingwei et al. (2015), Lawhern et al. (2018), Hartmann et al. (2018); Yang et al. (2015) Tang et al. (2017)
		Emotional EEG	Frydenlund and Rudzicz (2015)	Zhang et al. (2018b)	Li et al. (2016a), Liu et al. (2017b), Wang et al. (2018), Li et al. (2017a); Wang et al. (2016b)
		Mental Disease EEG	Yuan et al. (2018)	Talathi (2017), Ruffini et al. (2016)	Ullah et al. (2018), Acharya et al. (2018b), Acharya et al. (2018a), Morabito et al. (2016), Schirrmeister et al. (2017), Hosseini et al. (2017c), Johansen et al. (2016); Ansari et al. (2018) Hosseini et al. (2017a)
		Data Augmentation			
		Others	Teo et al. (2017), Reddy and Behera (2016), Yepes et al. (2017)	Shang et al. (2017)	Behncke et al. (2018), Hung et al. (2017), Baltatzis et al. (2017), Stober et al. (2014a), Shang et al. (2017), Völker et al. (2018), Hernández et al. (2018), Almogbel et al. (2018); Putten et al. (2018), Hajinoroozi et al. (2015b), Sternin et al. (2015); Chu et al. (2017)

(Continued)

Table 4.1: (*Continued*)

Brain Signals		Deep Learning Models — Representative Models				
		AE (D-AE)	RBM (D-RBM)	DBN-AE	DBN	DBN-RBM
Invasive						
Non-invasive Signals — EEG — Spontaneous EEG	Sleep EEG					Zhang et al. (2016a); Tan et al. (2015a)
	MI EEG	Li et al. (2015a), Redkar (2015), Zhang et al. (2017c)		Li and Cichocki (2014)	Ren and Wu (2014); Kumar et al. (2016), Lu et al. (2017)	
	Emotional EEG	Chai et al. (2016), Liu et al. (2016)	Zheng and Lu (2015), Zheng et al. (2015), Jia et al. (2014)	Xu and Plataniotis (2016a)	Jia et al. (2014), Xu and Plataniotis (2016a), Li et al. (2015b), Xu and Plataniotis (2016b), Zheng et al. (2014); Li et al. (2013)	
	Mental Disease EEG	Yuan et al. (2017), Lin et al. (2016), Morabito et al. (2017), Wen and Zhang (2018)		Page et al. (2014)	Zhao and He (2014); Turner et al. (2014)	
Data Augmentation						
	Others	Yin and Zhang (2017), Du et al. (2017)		Narejo et al. (2016)	Hajinoroozi et al. (2015a), Narejo et al. (2016), San et al. (2016); Li et al. (2016b)	

Table 4.1: (*Continued*)

Brain Signals				Deep Learning Models				
				Generative Models		LSTM + CNN	Hybrid Models	
				VAE	GAN		Repre + Discri	Others
Invasive			Sleep EEG			Supratak et al. (2017), Biswal et al. (2017)	Antoniades et al. (2018), Hosseini et al. (2017a), Xie et al. (2018), Fraiwan and Lweesy (2017)	Manzano et al. (2017), Dong et al. (2018)
Non-invasive Signals	EEG	Spont-aneous EEG	MI EEG	Dai et al. (2019)		Tan et al. (2017), Zhang et al. (2018h)	Tabar and Halici (2016), Duan et al. (2016)	Nurse et al. (2015), Zhang et al. (2018d), Wang et al. (2017), Sakhavi et al. (2015), Zhang et al. (2018i)
			Emotional EEG			Mioranda-Correa and Patras (2018)	Kawde and Verma (2017), Gao et al. (2015b), Yin et al. (2017)	Alhagry et al. (2017)
			Mental Disease EEG			Shah et al. (2017)	Hosseini et al. (2017b,a), Golmohammadi et al. (2017c), Al-kaysi et al. (2015)	
			Data Augmentation		Abdelfattah et al. (2018), Dai et al. (2019), Palazzo et al. (2017), Kavasidis et al. (2017)			
			Others			Hajinoroozi et al. (2015b)	Stober et al. (2015), Chai et al. (2017), Bashivan et al. (2015)	Zhang et al. (2018c)

(*Continued*)

Table 4.1: (*Continued*)

Brain Signals					Deep Learning Models — Discriminative Models		
					MLP	RNN	CNN
Non-invasive Signals	EEG	EP	ERP	VEP	Koike-Akino et al. (2016), Kawasaki et al. (2015)	Spampinato et al. (2017), Kavasidis et al. (2017)	Spampinato et al. (2017), Lawhern et al. (2018), Liu et al. (2018b); Hajinoroozi et al. (2015b), Hajinoroozi et al. (2015b); Sarkar et al. (2016), Cecotti and Graser (2011)
				RSVP	Mao et al. (2014), Mao (2016)		Manor and Geva (2015), Cecotti (2017), Solon et al. (2017), Hajinoroozi et al. (2017), Mao et al. (2017); Manor et al. (2016), Lin et al. (2017); Mao (2016), Gordon et al. (2017); Yoon et al. (2018), Shamwell et al. (2016); Cecotti et al. (2014)
				AEP			Carabez et al. (2017); Sarkar et al. (2016), Cecotti and Graser (2011); Stober et al. (2014b)
			SSEP				
			SSVEP		HACHEM et al. (2014)	Thomas et al. (2017)	Kwak et al. (2017), Waytowich et al. (2018), Thomas et al. (2017); Aznan et al. (2018), Tu et al. (2018)
	fNIRS				Naseer et al. (2016), Huve et al. (2017), Huve et al. (2018), Chiarelli et al. (2018), Hennrich et al. (2015)		Huve et al. (2017)
	fMRI				Koyamada et al. (2015), Shen et al. (2019)		Cichy et al. (2016), Jingwei et al. (2015), Havaei et al. (2017), Shreyas and Pankajakshan (2017), Hosseini et al. (2017c), Sarraf and Tofighi (2016), Li et al. (2014); Tu et al. (2018)
	EOG						Wang et al. (2016b); Sternin et al. (2015), Fernández-Varela et al. (2018)
	MEG						Garg et al. (2017); Cichy et al. (2016)

Table 4.1: (*Continued*)

Brain Signals					Deep Learning Models — Representative Models		DBN	
					AE (D-AE)	RBM (D-RBM)	DBN-AE	DBN-RBM
Non-invasive Signals	EEG	EP	ERP	VEP	Gao et al. (2015a)			Liu et al. (2017a); Ma et al. (2017), Sarkar et al. (2016)
				RSVP			Vařeka and Mautner (2017)	
				AEP				Sarkar et al. (2016)
				SSVEP				
			SSEP					
	fNIRS						Kulasingham et al. (2016)	Kulasingham et al. (2016)
	fMRI				Suk et al. (2016)		Suk et al. (2015)	Plis et al. (2014), Suk et al. (2015), Ortiz et al. (2016); Suhaimi et al. (2015)
	EOG				Du et al. (2017), Liu et al. (2016)			Xia et al. (2015)
	MEG				Shu and Fyshe (2013)			

(*Continued*)

Table 4.1: (*Continued*)

Brain Signals					Deep Learning Models				
					Generative Models		Hybrid Models		
					VAE	GAN	LSTM + CNN	Repre + Discri	Others
Non-invasive Signals	EEG	EP	ERP	VEP			Maddula et al. (2017), Bashivan et al. (2016b), Bashivan et al. (2016a)	Zheng et al. (2015), Shanbhag et al. (2017), Zheng and Lu (2015), Manor et al. (2016); Mao (2016)	Cecotti et al. (2014)
				RSVP					
				AEP					
			SSEP	SSVEP			Attia et al. (2018)	Pérez-Benítez et al. (2018) Hiroyasu et al. (2014)	
	fNIRS								
	fMRI					Seeliger et al. (2018), Han et al. (2018), Zhang et al. (2018d), Shen et al. (2019)		Hu et al. (2016)	
	EOG								
	MEG							Hasasneh et al. (2018)	

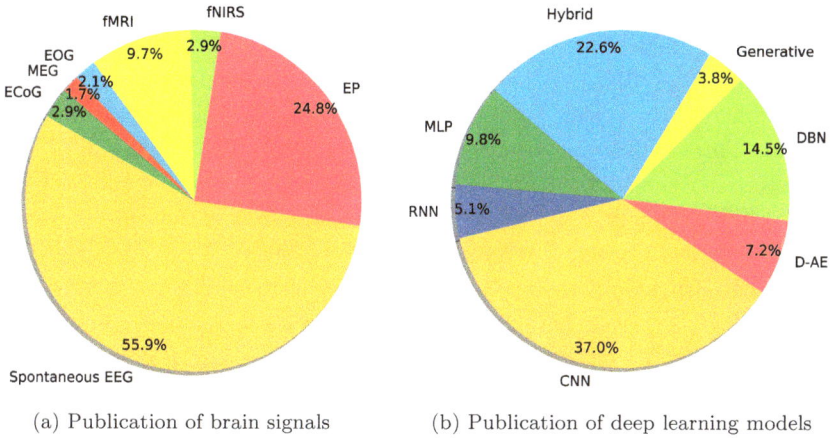

(a) Publication of brain signals   (b) Publication of deep learning models

Fig. 4.1: Illustration of the publication proportions for crucial brain signals and deep learning models.

(denoted as "Repre + Discri"), and others. Figure 4.1 presents the publication proportions for crucial brain signals and deep learning models.

4.7.1 *Discussions on Brain Signals*

Our research summarized above reveals that studies on noninvasive signals dominate the brain signal research. Among the 238 publications surveyed, only seven focused on invasive brain signal research, and most of those worked on ECoG rather than intracortical signals. One important reason for this is that the invasive brain research makes higher demands on the hardware and experimental environments. For example, the collection of ECoG signals requires a volunteer patient and a surgeon capable of performing a craniotomy, which disqualifies most researchers. Furthermore, there are few public data sets of invasive brain signals; invasive data is therefore simply inaccessible to most. In terms of the classification of invasive signals, CNN-related algorithms often display a higher capability of recognizing the pulse within cortex neurons.

Among the studies on the noninvasive signals, the number focusing on EEG alone is far greater than the sum of those on all of the other brain signal paradigms combined (fNIRS, fMRI, EOG, and MEG). Among those, about 70% of the EEG papers consider spontaneous EEG (133 publications). For increased understanding, we split the studies on spontaneous EEG into several categories: sleeping, motor imagery, emotional, mental disease, data augmentation, and other. First, classification of the sleep EEG mainly depended on the discriminative and the hybrid models. Among the 19 studies on sleep stage classification, six employed CNN and the modified CNN models independently, whereas two papers adopted RNN models. Three hybrid models were a combination of CNN and RNN.

In terms of the research on MI EEG (30 publications), independent CNN and CNN-based hybrid models were widely used. As for the representative models, DBN-RBM was often applied to capture the latent features from the MI EEG signals. Twenty-five publications were related to spontaneous emotional EEG. Over half employed representative models (such as D-AE, D-RBM, and especially DBN-RBM) for unsupervised feature learning. Most affective state recognition work recognized the user's emotion as positive, neutral, or negative. Some researchers took the further step of classifying the valence and the degree of arousal, which is more complex and challenging.

The research into mental disease diagnosis is promising and attractive. The majority of this research focuses on the detection of epileptic seizure and Alzheimer's disease. Since the detection is a binary classification problem, many studies have achieved high accuracies of above 90%. The standard CNN model and the D-AE are prevalent in this area, one possible reason for which is that CNN and AE are the best-known and most-effective deep learning models for classification and dimensionality reduction. Several publications have implemented GAN-based data augmentation. About 30 studies investigated other spontaneous EEG such as driving fatigue, audio/visual stimuli impact, cognitive/mental load, and eye state detection. These studies extensively applied standard CNN models and variants.

Apart from spontaneous EEG, evoked potentials also drew much attention. On the one hand, in ERP, studies using VEP and its subcategory RSVP are common because visual stimuli, compared to other stimuli, are more easily conducted have more real-world applications (e.g., the P300 speller can be used for brain typing). Whereas for VEP (21 publications), 11 studies applied discriminative models and six works adopted hybrid models, for RSVP, the sole CNN model dominated the algorithms. Apart from these, five studies focused on the analysis of AEP signals, while among the steady state-related research, only SSVEP has been studied using deep learning models. Most of these papers only applied discriminative models to the recognition of the target image.

Furthermore, beyond the diverse use of EEG diagrams, a wide range of studies targeted fNIRS and fMRI. However, fNIRS images were rarely studied using deep learning, and the major studies in this area employed simple MLP models. We believe fNIRS warrants further research for its high portability and low cost. As for fMRI, 23 papers proposed proposed deep learning models for the classification. The CNN model was widely used for its outstanding performance in feature learning from images. There were also several studies into image reconstruction based on fMRI signals. One reason why fMRI is so hot is that, although the fMRI equipment is expensive, several public data sets are available online. Nowadays, EOG is mainly regarded as noise rather than as a useful signal. However, it can enable an immobilized individual to communicate with the outside world by registering their eye movement. The MEG signals are mainly used in the clinical area; thus, we found very few related studies. The sparse AE and CNN algorithms make positive contributions to the feature refining and classification by MEG.

4.7.2 *Discussions on Deep Learning Models*

Our analysis shows that discriminative models appeared most frequently in the summarized publications — which is reasonable at a high level because a large proportion of brain signal issues can be regarded as a classification problem. Another observation is that CNN represents more than 70% of the discriminative models, for which we provide reasons as follow.

First, the design of CNN is powerful enough to extract the latent discriminative features and spatial dependencies from the EEG signals for classification. As a result, CNN structures are adopted for classification in some studies and for feature engineering in others. Second, CNN has achieved great success in some research areas (e.g., computer vision), which has made it both extremely well known and feasible (public codes). Brain signal researchers are thus more likely to understand CNN and to be able apply it in their work. Third, some brain diagrams (e.g., fMRI) naturally form 2D images conducive to processing by CNN, while other 1D signals (e.g., EEG) are easily converted into 2D images for further analysis by CNN. Here, we provide several methods for converting 1D EEG signals (with multiple channels) to a 2D matrix: (1) convert each time point[1] to a 2D image; (2) convert a segment into a 2D matrix. For the first situation,

---

[1] One time point represents one sampling point. For example, we would have 100 time points at a sampling rate of 100 Hz.

suppose we have 32 channels and we can collect 32 elements (each element corresponding to a channel) at each time point. As described in Li *et al.* (2016a), the collected 32 elements could be converted into a 2D image based on their spatial organization. For the second situation, suppose we have 32 channels and the segment contains 100 time points. The collected data can be arranged as a matrix with the shape of $(32, 100)$ where each row and column refers to a specific channel and time point, respectively.

Fourth, there are a lot of variants of CNN which are suitable in a wide range of brain signal scenarios. For example, single-channel EEG signals can be processed by 1D CNN. In terms of RNN, it was adopted in only about 20% of the discriminative model-based papers, which is far fewer than we had expected since RNN has proved to be powerful at temporal feature learning. One possible reason for this is that processing a long sequence by RNN is time-consuming and the EEG signals typically form long sequences. For example, the sleep signals are usually sliced into 30-s segments, each of which has 3000 time points given a 100 Hz sampling rate. For a sequence containing 3000 elements, based on our preliminary experiments, RNN would take more than twentyfold the training time of CNN. Moreover, MLP is not popular due to its inferior effectiveness (e.g., nonlinear ability) compared to the other algorithms.

Among representative models, DBN, especially DBN-RBM, is the most popular model for feature extraction. DBN is widely used in neurological research for two reasons: (1) it learns the generative parameters that reveal the relationship of variables in neighboring layers efficiently; (2) it makes calculating the values of latent variables in each hidden layer straightforward (Deng, 2012). However, most work employing the DBN-RBM model was published prior to 2016, indicating that DBN is no longer popular. It can be inferred that researchers preferred to use DBN for feature learning followed by a nondeep learning classifier before 2016; but recently, an increasing number of studies have begun to adopt CNN or hybrid models for both feature learning and classification.

Moreover, generative models are rarely employed independently. The GAN- and VAE-based data augmentation and image reconstruction are mainly focused on fMRI and EEG signals. It has been demonstrated that a trained classifier will perform more competitively after data augmentation; therefore, this presents a promising prospect future research.

Last but not least, 53 publications proposed hybrid models for brain signal studies. Of these, combinations of RNN and CNN made up about one-fifth. Since RNN and CNN are viewed as equally excellent for spatiotemporal feature extraction, it is natural to combine them for both temporal and spatial feature learning. Another type of hybrid models is the combination of representative and discriminative models. This is a logical combination; the former is employed to refine feature learning, while the latter is employed for classification. There are 28 publications which among them nearly covered every type of brain signal proposed for this type of hybrid deep learning model. The adopted representative models were mostly AE or DBN-RBM; meanwhile, the adopted discriminative models were mostly CNN. Apart from those, 12 papers proposed other types of hybrid models such as two discriminative models. For example, several studies proposed a combination of CNN and MLP in which the CNN structure was used to extract spatial features, which were then fed into an MLP for classification.

Chapter 5

Deep Learning-Based BCI Applications

Deep learning models have contributed to various of BCI applications as summarized in Table 5.1. Studies focusing on signal classification without application background are not listed. In the table, "S-EEG," "MD EEG," and "E-EEG," respectively, denote sleep EEG, mental disease EEG, and emotional EEG. "EEG" alone refers to the other subcategories of spontaneous EEG. In the performance column, "N/A," "sen," "spe," "aro," "val," "dom," and "like" denote not found, sensitivity, specificity, arousal, valence, dominance, and liking, respectively. For each application scenario, the literature are organized by signal types and deep learning models.

5.1 Health Care

In the field of health care, the deep learning-based BCI systems mainly work on the detection and diagnosis of mental disorders such as sleep disorders, Alzheimer's disease, epileptic seizure, and other conditions. For sleep disorder detection, most studies focus on the sleep stage detection based on spontaneous sleep EEGs. In this context, the researchers do not need to recruit sleep-disordered patients because the sleep EEG signals can easily be collected from healthy individuals. In terms of the algorithms used, as shown in Table 5.1, DBN-RBM and CNN are widely adopted for feature engineering and classification. Ruffini *et al.* (2016) go one step further by applying them to detect REM sleep behavior disorder (RBD), which may cause neurodegenerative diseases such as Parkinson's disease. They achieved an average accuracy of 85% in differentiating those subjects with RBD from the healthy controls.

Table 5.1: Summary of deep learning-based BCI applications. A "local" data set is either a private or unavailable data set; the public data sets (with links) will be introduced in Section 5.9.

BCI Applications	Reference	Signals	Deep Learning Models	Data Set	Performance
Sleeping Quality Evaluation	Shahin et al. (2017)	S-EEG	MLP	University Hospital in Berlin	0.9
	Biswal et al. (2017)	S-EEG	RNN	Local	0.8576
	Ruffini et al. (2016)	S-EEG	RNN	Local	0.85
	Vilamala et al. (2017)	S-EEG	CNN	Sleep-EDF	0.86
	Tsinalis et al. (2016)	S-EEG	CNN	Sleep-EDF	0.82
	Sors et al. (2018)	S-EEG	CNN	SHHS	0.87
	Chambon et al. (2018)	S-EEG	Multi-view CNN	MASS session 3	N/A
	Manzano et al. (2017)	S-EEG	CNN + MLP	Sleep-EDF	0.732
	Fraiwan and Lweesy (2017)	S-EEG	DBN-AE + MLP	Local	0.804
	Tan et al. (2015a)	S-EEG	DBN-RBM	Local	0.9278 (F1)
	Zhang et al. (2016a)	S-EEG	DBN + voting	UCD	0.9131
	Fernández-Varela et al. (2018)	EEG, EOG	CNN	SHHS	0.9 (F1)
	Supratak et al. (2017)	EEG, EOG	CNN + LSTM	MASS/ Sleep-EDF	0.862/0.82
Health Care	Xia et al. (2015)	EOG	DBN-RBM + HMM	Sleep-EDF	0.833
AD Detection	Morabito et al. (2016)	MD EEG	CNN	Local	0.82
	Zhao and He (2014)	MD EEG	DBN-RBM	Local	0.92
	Suk et al. (2015)	fMRI	DBN-AE; DBN-RBM	ADNI	0.979; 0.954
	Sarraf and Tofighi (2016)	fMRI	CNN	ADNI	0.9685
	Li et al. (2014)	fMRI	CNN + LR	ADNI	0.9192
	Hu et al. (2016)	fMRI	D-AE + MLP	ADNI	0.875
	Ortiz et al. (2016)	fMRI, PET	DBN-RBM + SVM	ADNI	0.9

Table 5.1: (*Continued*)

BCI Applications	Reference	Signals	Deep Learning Models	Data Set	Performance
Seizure Detection	Kiral-Kornek et al. (2018)	EcoG	MLP	Local	Sen: 0.69
	Hosseini et al. (2017a)	EEG, EcoG	CNN	Local	0.96
	Yuan et al. (2018)	MD EEG	Attention-MLP	CHB-MIT	0.9661
	Tsiouris et al. (2018)	MD EEG	LSTM	CHB-MIT	>0.99
	Talathi (2017)	MD EEG	GRU	BUD	0.996
	Acharya et al. (2018a)	MD EEG	CNN	UBD	0.8867
	Schirrmeister et al. (2017)	MD EEG	CNN	TUH	0.854
	Hosseini et al. (2017c)	MD EEG	CNN	Local	N/A
	Johansen et al. (2016)	MD EEG	CNN	Local	0.947 (AUC)
	Ansari et al. (2018)	MD EEG	CNN + RF	Local	0.77
	Ullah et al. (2018)	MD EEG	CNN + voting	UBD	0.954
Health Care	Wen and Zhang (2018)	MD EEG	AE	Local	0.92
	Lin et al. (2016)	MD EEG	D-AE	UBD	0.96
	Yuan et al. (2017)	MD EEG	D-AE + SVM	CHB-MIT	0.95
	Page et al. (2014)	MD EEG	DBN-AE + LR	N/A	0.8 ~ 0.9
	Turner et al. (2014)	MD EEG	DBN-RBM + LR	Local	N/A
	Hosseini et al. (2017b)	MD EEG	D-AE + MLP	Local	0.94
	Golmohammadi et al. (2017c)	MD EEG	RNN+CNN	TUH	Sen: 0.3083; Spe: 0.9686
	Shah et al. (2017)	MD EEG	CNN+ LSTM	TUH	Sen: 0.39; Spe: 0.9037

(*Continued*)

Table 5.1: (*Continued*)

BCI Applications		Reference	Signals	Deep Learning Models	Data Set	Performance
	Others:					
	Interictal Epileptic Discharge (IED)	Antoniades et al. (2016)	EcoG	CNN	Local	0.8751
		Antoniades et al. (2018)	EEG, EcoG	AE + CNN	Local	0.68
	Creutzfeldt-Jakob Disease (CJD)	Morabito et al. (2017)	MD EEG	D-AE	Local	0.81 ~ 0.83
	Depression	Acharya et al. (2018b)	MD EEG	CNN	Local	0.935 ~ 0.9596
		Al-kaysi et al. (2015)	MD EEG	DBN-RBM + MLP	Local	0.695
Health Care	Brain Tumor	Marc Moreno (2017)	fMRI	CNN	BRATS	0.88 (F1)
		Shreyas and Pankajakshan (2017)	fMRI	CNN	BRATS	0.83
		Havaei et al. (2017)	fMRI	Muli-scale CNN	BRATS	0.88 (F1)
	Schizophrenia	Plis et al. (2014)	fMRI	DBN-RBM	Combined	>0.9 (F1)
		Chu et al. (2017)	fMRI	CNN + RF + Voting	Local	0.816, 0.967, 0.992
	Mild Cognitive Impairment (MCI)	Suk et al. (2016)	fMRI	AE + SVM	ADNI2	0.7258
	Cardiac Detection	Garg et al. (2017)	MEG	CNN	Local	Sen: 0.85, Spe: 0.97
		Hasasneh et al. (2018)	MEG	CNN + MLP	Local	0.944
Smart Environment	Robot Control	Behncke et al. (2018)	EEG	CNN	Local	0.75
	Smart Home	Zhang et al. (2017b)	MI EEG	RNN	EEGMMI	0.9553
	Exoskeleton Control	Kwak et al. (2017)	SSVEP	CNN	Local	0.9403
		Huve et al. (2018)	fNIRS	MLP	Local	0.82

Table 5.1: (*Continued*)

BCI Applications		Reference	Signals	Deep Learning Models	Data Set	Performance
Communication		Zhang *et al.* (2018h)	MI EEG	LSTM+CNN+AE	Local	0.9452
		Kawasaki *et al.* (2015)	VEP	MLP	Local	0.908
		Cecotti and Graser (2011)	VEP	CNN	The third BCI competition, data set II	0.945
		Liu *et al.* (2018b)	VEP	CNN	The third BCI competition, data set II	0.92 ~ 0.96
		Cecotti and Graser (2011)	VEP	CNN + Voting	The third BCI competition, data set II	0.955
		Maddula *et al.* (2017)	VEP	RCNN	Local	0.65 ~ 0.76
Security	Identification	Zhang *et al.* (2018g)	MI-EEG	Attention-based RNN	EEGMMI + local	0.9882
	Authentication	Koike-Akino *et al.* (2016)	VEP	MLP	Local	0.976
		Mao *et al.* (2017)	RSVP	CNN	Local	0.97
		Zhang *et al.* (2019b)	MI EEG	Hybrid	EEGMMI + local	0.984
Affective Computing		Frydenlund and Rudzicz (2015)	E-EEG	MLP	DEAP	N/A
		Zhang *et al.* (2018b)	E-EEG	RNN	SEED	0.895
		Hiroyasu *et al.* (2014)	E-EEG	CNN	SEED	0.882
		Liu *et al.* (2017b)	E-EEG	CNN	Local	0.82
		Li *et al.* (2016a)	E-EEG	Hierarchical CNN	SEED	0.882
		Chai *et al.* (2016)	E-EEG	AE	SEED	0.818
		Xu and Plataniotis (2016a)	E-EEG	DBN-AE, DBN-RBM	DEAP	>0.86 (F1)
		Jia *et al.* (2014)	E-EEG	DBN-RBM	DEAP	0.8 ~
		Li *et al.* (2015b)	E-EEG	DBN-RBM	DEAP	0.85 (AUC) Aro:0.642, Val:0.584, Dom 0.658
		Xu and Plataniotis (2016b)	E-EEG	DBN-RBM	DEAP	Aro:0.6984, Val:0.6688, Lik: 0.7539

(*Continued*)

Table 5.1: (Continued)

BCI Applications	Reference	Signals	Deep Learning Models	Data Set	Performance
	Zheng et al. (2014)	E-EEG	DBN-RBM + HMM	Local	0.8762
	Zheng and Lu (2015) and Zheng et al. (2015)	E-EEG	DBN-RBM + MLP	SEED	0.8608
	Gao et al. (2015b)	E-EEG	DBN-RBM + MLP	Local	0.684
	Yin et al. (2017)	E-EEG	Multi-view D-AE + MLP	DEAP	Aro: 0.7719; Val: 0.7617
Affective Computing	Mioranda-Correa and Patras (2018)	E-EEG	RNN + CNN	AMIGOS	<0.7
	Alhagry et al. (2017)	E-EEG	LSTM + MLP	DEAP	Aro: 0.8565, Val: 0.8545, Lik: 0.8799
	Liu et al. (2016)	EEG, EOG	AE	SEED, DEAP	0.9101, 0.8325
	Kawde and Verma (2017)	EEG, EOG	DBN-RBM	DEAP	Aro: 0.7033; Val: 0.7828; Dom: 0.7016
	Hung et al. (2017)	EEG	CNN	Local	0.572 (RMSE)
	Almogbel et al. (2018)	EEG	CNN	Local	0.9531
	Hajinoroozi et al. (2015b,b)	EEG	CNN	Local	0.8294
Drive Fatigue Detection	Hajinoroozi et al. (2015a)	EEG	DBN-RBM	Local	0.85
	San et al. (2016)	EEG	DBN-RBM + SVM	Local	0.7392
	Chai et al. (2017)	EEG	DBN + MLP	Local	0.931
	HACHEM et al. (2014)	SSVEP	MLP	Local	0.75
	Du et al. (2017)	EEG, EOG	D-AE + SVM	Local	0.094 (RMSE)
	Yin and Zhang (2017)	EEG	D-AE	Local	0.9584
	Bashivan et al. (2015)	EEG	DBN-RBM	Local	0.92
	Li et al. (2016b)	EEG	DBN-RBM	Local	0.9886
Mental Load Measurement	Bashivan et al. (2016b)	EEG	R-CNN	Local	0.9111
	Bashivan et al. (2016a)	EEG	DBN + MLP	Local	N/A
	Naseer et al. (2016)	fNIRS	MLP	Local	0.963
	Hennrich et al. (2015)	fNIRS	MLP	Local	0.641

Table 5.1: (*Continued*)

BCI Applications		Reference	Signals	Deep Learning Models	Data Set	Performance
	Finger Trajector	Xie et al. (2018)	EcoG	RNN+CNN	BCI Competition IV	N/A
	School Bullying	Baltatzis et al. (2017)	EEG	CNN	Local	0.937
	Music Detection	Stober et al. (2014a)	EEG	CNN	Local	0.776
		Stober et al. (2015)	EEG	AE + CNN	Open MIIR	0.27 for 12-class
		Stober et al. (2014b)	EEG	CNN	Local	0.244
		Sternin et al. (2015)	EEG, EOG	CNN	Local	0.75
	Number Choosing	Waytowich et al. (2018)	SSVEP	CNN	Local	0.8
	Visual Object Recognition	Cichy et al. (2016)	fMRI, MEG	CNN	N/A	N/A
		Manor and Geva (2015)	RSVP	CNN	Local	0.75
		Cecotti (2017)	RSVP	CNN	Local	0.897 (AUC)
		Hajinoroozi et al. (2017)	RSVP	CNN	Local	0.7242 (AUC)
		Shamwell et al. (2016)	RSVP	CNN	Local	0.7252 (AUC)
Other Applications		Pérez-Benítez et al. (2018)	SSVEP	AE	Local	0.9778
	Guilty Knowledge Test	Kulasingham et al. (2016)	SSVEP	DBN-RBM; DBN-AE	Local	0.869; 0.8601
	Concealed Information Test	Liu et al. (2017a)	EEG	DBN-RBM	Local	0.973
	Flanker Task	Völker et al. (2018)	EEG	CNN	Local	0.841
	Eye State	Narejo et al. (2016)	EEG	DBN-RBM	UCI	0.989
		Reddy and Behera (2016)	EEG	MLP	Local	0.975
	User Preference	Teo et al. (2017)	EEG	MLP	Local	0.6399
	Emergency Braking	Hernández et al. (2018)	EEG	CNN	Local	0.718
	Gender Detection	Putten et al. (2018)	EEG	CNN	Local	0.81
		Hiroyasu et al. (2014)	fNIRS	D-AE + MLP	Local	0.81

fMRI is widely applied in the diagnosis of Alzheimer's dis-ease; its characteristically high spatial resolution yields a diagnosis accuracy rate of above 90% in several studies. Another factor contributing to its competitive performance is that the diagnosis is a binary classification task. scenario. Apart from that, several publications diagnose AD based on spontaneous EEG (Morabito *et al.*, 2016; Zhao and He, 2014). Diagnosis of epileptic seizure also drew interest. Seizure detection is based mainly on spontaneous EEG and a few ECoG signals. Popular deep learning models in this scenario contain both independent CNN and RNN, as well as hybrid models of combined RNN and CNN. Some models integrate deep learning models for feature extraction and use a traditional classifier for detection (Page *et al.*, 2014; Turner *et al.*, 2014). For example, Yuan *et al.* (2017) applied a D-AE in feature engineering followed by an SVM for seizure diagnosis. Ullah *et al.* (2018) adopted the voting for post-processing, in which several CNN classifiers are proposed and the final prediction was the majority result.

Many other healthcare issues can be identified using BCI systems. The cardiac artifacts in MEG can be automatically detected by deep learning models (Garg *et al.*, 2017; Hasasneh *et al.*, 2018). Several modified CNN structures are proposed to detect brain tumor based on fMRI from the public BRATS data set (Havaei *et al.*, 2017; Shreyas and Pankajakshan, 2017). And researchers have demonstrated the effectiveness of deep learning models in the detection of a wide range of mental diseases such as depression (Acharya *et al.*, 2018b), interictal epileptic discharge (IED) (Antoniades *et al.*, 2016), schizophrenia (Plis *et al.*, 2014), Creutzfeldt-Jakob Disease (CJD) (Morabito *et al.*, 2017), and Mild Cognitive Impairment (MCI) (Suk *et al.*, 2016).

5.2   Smart Environments

The smart environment is a promising scenario for brain signal application. With the development of the Internet of Things (IoT), an increasing number of smart environments can be connected to the brain signal system. For example, an assistive robot used in a smart home (Zhang *et al.*, 2017b, 2018i), can be controlled by the brain signals of its inhabitants. Behncke *et al.* (2018) and Huve *et al.* (2018) investigated the robot control problem based on the visual stimulated spontaneous EEG and fNIRS signals. The brain-controlled exoskeleton could assist individuals with damage to the lower limb motor neurons in walking and daily activities (Kwak *et al.*, 2017). In the future, the research on mind-controlled appliances may benefit the elderly and those with disabilities in smart homes and smart hospitals.

5.3 Communication

One of the greatest benefits of brain signal research, compared with other human–machine interface techniques, is that brain signals enable a patient who has lost most motor abilities and is unable to speak to continue to verbally communicate. Deep learning technology has improved the efficiency of brain signal-based communication. One popular diagram, which enables an individual to type without any motor system involvement, is the P300 speller, which can convert the user's intent into text (Kawasaki *et al.*, 2015). Powerful deep learning models empower the BCI systems to separate the P300 segment, which contains the brain speech signal, from the non-P300 segment (Cecotti and Graser, 2011). At a higher level, the representative deep learning models can rapidly detect each character as the as the user's focus runs through a series and prints them sequentially on a screen, allowing them to chat with others (Cecotti and Graser, 2011; Liu *et al.*, 2018b; Maddula *et al.*, 2017). Additionally, Zhang *et al.* (2018h) proposed a hybrid model of combined RNN, CNN, and AE that extracts the informative features from MI EEGs to recognize each letter the user wants to speak.

5.4 Security

Brain signals can also be used to handle security issues such as identification (or recognition) and authentication (or verification). For the former, multiclass classification is performed to determine a person's identity (Zhang *et al.*, 2018g). The latter employs binary classification to test authorization (Zhang *et al.*, 2019b).

The majority of the existing biometric identification/authentication systems rely on individuals' intrinsic physiological features such as facial structure, iris, retina, voice, and fingerprint (Zhang *et al.*, 2018g). They are vulnerable to various means of attack including anti-surveillance prosthetic masks, contact lenses, vocoders, and fingerprint films. EEG-based biometric identification is a promising alternative given that it is highly resilient against spoofing attacks — individual's EEG signals are virtually impossible for an imposter to mimic. Koike-Akino *et al.* (2016) have adopted deep neural networks to identify a user's ID based on VEP signals; Mao *et al.* (2017) applied CNN model to person identification based on RSVP signals; Zhang *et al.* (2018g) proposed an attention-based LSTM model and evaluated it over both public and local data sets. And EEG signals have also been combined with gait information in a hybrid deep learning model in order to build a dual-authentication system (Zhang *et al.*, 2019b).

5.5   Affective Computing

The affective states of a user provide critical information for many applications, such as retrieval of personalized information (e.g., multi-media content) or intelligent human-computer interface design (Xu and Plataniotis, 2016a). Recent research illustrates that deep learning models can enhance the performance in affective computing. The most widely used circumplex model posits that the emotions are distributed in two dimensions: arousal and valence. Arousal refers to the intensity of the emotional stimuli or strength of the emotion; valence refers to value associated with the stimulus by the person experiencing the emotion. In some other models, the dominance and liking are also deployed.

A few researchers (Li *et al.*, 2016a; Liu *et al.*, 2017b; Wang *et al.*, 2018) have attempted to classify users' emotional states into two (positive/negative) or three categories (positive, neutral, and negative) based on EEG signals using deep learning algorithms such as CNN and its variants (Frydenlund and Rudzicz, 2015). DBN-RBM is the most representative deep learning model to have discovered the concealed features from emotional spontaneous EEG (Xu and Plataniotis, 2016a; Zheng and Lu, 2015). Xu and Plataniotis (2016a) applied DBN-RBM as a feature extractor method to classify affective states based on EEG.

Furthermore, a number of other researchers aim to recognize the positive/negative state of each specific emotional dimension. For example, Yin *et al.* (2017) employed an ensemble classifier of AE in order to recognize the user's affection. Each AE uses three hidden layers to filter out noise and to derive stable physiological feature representations. The proposed model was evaluated over the benchmark, DEAP data set and achieved classification accuracy for arousal and valence of 77.19% and 76.17%, respectively.

5.6   Driver Fatigue Detection

Drivers' ability to keep alert and maintain optimal performance will dramatically affect traffic safety (Almogbel *et al.*, 2018). EEG signals have proven useful in evaluating humans cognitive states in different contexts. Generally, a driver is regarded as in an alert state if their reaction time is less than 0.7 s and in fatigued state if it is over 2.1 s. Hajinoroozi *et al.* (2015a) proposed that driver fatigue could be detected from EEG signals by discovering the distinct corresponding features. They explored a DBN-based approach to dimension reduction.

Detecting driver fatigue is essential to prevent accidents caused by driver drowsiness and is feasible in practice. EEG collection equipment s affordable hardware that is available off the shelf and portable enough to be used in a car. From the technological perspective, deep learning models have greatly enhanced the performance of fatigue detection. As we summarized, drowsiness can be recognized from EEG signals with a high rate of accuracy (82–95%).

The future of driver fatigue detection lies in the self-driving scenario. In the most situation of self-driving situations (e.g., Automation level 3[1]), the human driver is expected to respond appropriately to a request to intervene, which means that the driver should remain alert. Therefore, we believe the application of brain signal-based driver fatigue detection will benefit the development of the self-driving car.

5.7 Mental Load Measurement

The EEG oscillations can be used to measure the mental workload level, which can in turn sustain decision making and strategy development in the context of human-machine interaction (Yin and Zhang, 2017). For example, an excessive mental workload on the human operator may result in performance degradation and cause catastrophic accidents (Parasuraman and Jiang, 2012).

Several researchers have been explored to this topic, utilizing both fNIRS signals and spontaneous EEG. Naseer *et al.* adopted an MLP algorithm for fNIRS-based binary mental task level classification (mental arithmetic and rest) which outperformed the traditional classifiers SVM, ANN k-NN, and achieved the highest accuracy at 96.3%. Bashivan *et al.* (2015) presented a statistical approach utilizing a DBN model, for the recognition of cognitive workload level based on single-trial EEG. Before applying the DBN, the authors manually extracted the wavelet entropy and band-specific power from three frequency bands (theta, alpha, and beta). Overall, these experiments demonstrated an overall accuracy of 92%.

---

[1]https://en.wikipedia.org/wiki/Self-driving_car. Accessed Oct. 15, 2018.

5.8   Other Applications

There are plenty of interesting scenarios beyond the above mentioned to which deep learning-based brain signal systems could be applied, such as recommender systems (Teo *et al.*, 2017) and emergency braking (Hernández *et al.*, 2018). One potential topic that seems particularly attractive is the recognition of a visual object, which could be used in both the guilty knowledge test (Kulasingham *et al.*, 2016) and the concealed information test (Liu *et al.*, 2017a). The neurons of the participant will produce a pulse when he/she is suddenly presented with an object similar to the one in question. Based on the theory, the visual target recognition is mainly used RSVP signals. Cecotti (2017) aimed to build a common model for target recognition, that would work for multiple subjects rather than only one, a specific subject.

Researchers have also investigated the possibility of distinguishing a subject's gender by fNIRS (Hiroyasu *et al.*, 2014) and spontaneous EEG (Putten *et al.*, 2018). Hiroyasu *et al.* (2014) adopted deep learning to recognize the gender of a subject based on cerebral blood flow. The results of the experiment suggest that regional cerebral blood flow rates differ between men and women. Putten *et al.* (2018) tried to detect sex-specific differences in brain rhythms by training a CNN model to assess the participant's gender while comparing sets of EEG recordings, which indicated that fast beta activity (20–25 Hz) is one of the most distinctive attributes.

5.9   Benchmark Data Sets

Collection of brain signals is both financially and temporally costly, for several reasons. First, it is hard to recruit a large set of volunteers for a study, especially when the focus is a specific group (e.g., patients with mental disease). Second, such a study may require tracking the group over a long span of time. For instance, sleep disorder generally last several months. Third, some medical monitoring equipment (e.g., fMRI device) is expensive and not readily available.

In this section, we extensively explore the benchmark data sets applicable to deep learning-based brain signal research. Table 5.2 provides 31 public data sets with download links that cover most brain signal types.

Table 5.2: The summary of public data sets for BCI systems. "# Sub," "# Cla," and "S-Rate" denote the number of subjects, number of classes, and sampling rates, respectively. "FM" denotes finger movement while "BCI-C" denotes the BCI competition. Data sets in the same BCI competition share the same download links.

Brain Signals		Name Link	# Sub	# Cla	S-Rate	# Channel
Invasive	FM EcoG	BCI-C IV[a], Data set IV	3	5	1000	48 ~ 64
	MI EcoG	BCI-C III[b], Data set I	1	2	1000	64
EEG	Sleeping EEG	Sleep-EDF[c]: Telemetry	22	6	100	2 EEG, 1 EOG, 1 EMG
		Sleep-EDF: Cassette	78	6	100, 1	2 EEG (100 Hz), 1 EOG (100 Hz), 1 EMG (1 Hz)
		MASS-1[d]	53	5	256	17/19 EEG, 2 EOG, 5 EMG
		MASS-2	19	6	256	19 EEG, 4 EOG, 1EMG
		MASS-3	62	5	256	20 EEG, 2 EOG, 3 EMG
		MASS-4	40	6	256	4 EEG, 4 EOG, 1 EMG
		MASS-5	26	6	256	20 EEG, 2 EOG, 3 EMG
		SHHS[e]	5804	N/A	125, 50	2 EEG (125 Hz), 1EOG (50 Hz), 1 EMG (125 Hz)
	Seizure EEG	CHB-MIT[f]	22	2	256	18
		TUH[g]	315	2	200	19
	MI EEG	EEGMMI[h]	109	4	160	64
		BCI-C II[i], Data set III	1	2	128	3
		BCI-C III, Data set III a	3	4	250	60
		BCI-C III, Data set III b	3	2	125	2
		BCI-C III, Data set IV a	5	2	1000	118
		BCI-C III, Data set IV b	1	2	1001	119
		BCI-C III, Data set IV c	1	2	1002	120
		BCI-C IV, Data set I	7	2	1000	64
		BCI-C IV, Data set II a	9	4	250	22 EEG, 3 EOG
		BCI-C IV, Data set II b	9	2	250	3 EEG, 3 EOG

(Continued)

Table 5.2: (*Continued*)

Brain Signals			Name Link	# Sub	# Cla	S-Rate	# Channel
EEG	Emotional EEG		AMIGOS[j]	40	4	128	14
			SEED[k]	15	3	200	62
			DEAP[l]	32	4	512	32
	Others EEG		Open MIIR[m]	10	12	512	64
	VEP		BCI-C II, Data set II b	1	36	240	64
			BCI-C III, Data set II	2	26	240	64
fMRI			ADNI[n]	202	3	N/A	N/A
			BRATS[o] 2013	65	4	N/A	N/A
MEG			BCI-C IV, Data set III	2	4	400	10

[a]http://www.bbci.de/competition/iv/. Accessed Dec. 15, 2020.
[b]http://www.bbci.de/competition/iii/. Accessed Dec. 15, 2020.
[c]https://physionet.org/physiobank/database/sleep-edfx/. Accessed Dec. 15, 2020.
[d]https://massdb.herokuapp.com/en/. Accessed Dec. 15, 2020.
[e]https://archive.physionet.org/pn3/shhpsgdb/. Accessed Sep. 2, 2021.
[f]https://archive.physionet.org/pn6/chbmit/. Accessed Sep. 2, 2021.
[g]https://www.isip.piconepress.com/projects/tuh_eeg/html/downloads.shtml. Accessed Dec. 15, 2020.
[h]https://archive.physionet.org/pn4/eegmmidb/. Accessed Sep. 2, 2021.
[i]http://www.bbci.de/competition/ii/. Accessed Dec. 15, 2020.
[j]http://www.eecs.qmul.ac.uk/mmv/datasets/amigos/readme.html. Accessed Dec. 15, 2020.
[k]http://bcmi.sjtu.edu.cn/~seed/download.html. Accessed Dec. 15, 2020.
[l]https://www.eecs.qmul.ac.uk/mmv/datasets/deap/. Accessed Dec. 15, 2020.
[m]https://owenlab.uwo.ca/research/the_openmiir_dataset.html. Accessed Dec. 15, 2020.
[n]http://adni.loni.usc.edu/data-samples/access-data/. Accessed Dec. 15, 2020.
[o]https://www.med.upenn.edu/sbia/brats2018/data.html. Accessed Dec. 15, 2020.

BCI competition IV (BCI-C IV) contains five data sets via the same link. For clarity, we include the number of subjects, the number of classes (categories), the sampling rate, and the number of channels in each data set. In the "# Channel" column, the default EEG . (Some data sets contain additional biometric signals, e.g., ECG, but for our purposes, only channels related to brain signals are listed here.)

5.10   Discussions

In order to make a closer assessment of the recent advances in deep learning-based brain signal analysis, it is necessary to analyze the signal acquisition methods and the deep learning algorithms in terms of their performance. In some cases, studies which have adopted the same deep architecture for working on the same data set but have had unequal results, possibly caused by the use of different pre-processing methods and hyperparameter settings.

To begin with, the most appealing and hottest field is the application of brain signal analysis in health care. For sleep quality evaluation, the predominant brain signals used are spontaneous EEG phenomena, which are recorded while the patient is sleeps. In some studies, the EOG signals are treated as auxiliary information for improved accuracy in the detection of the eye movement stage. The single RNN and CNN models seem to have a powerful discriminative feature learning ability that delivers a capable performance. Generally, most of the deep learning algorithms achieve an accuracy of above 85% in the context of multiple stages, upon which the combined hybrid models (e.g., CNN integrated with LSTM) can make only incremental improvements.

One key method of detecting Alzheimer's disease is by the analysis of brain signal performed by measuring the functions of specific brain regions. In detail, the diagnosis can be made using spontaneous EEG signals or fMRI images. For MD EEG, DBN is assumed to outperform CNN since the EEG signals contain more temporal than spatial information. As for the fMRI images, CNN has great advantages in the grid-arranged spatial information learning, which gives it a very comprehensive classification accuracy (above 90%). As for epileptic seizure, diagnosis is generally based on EEG signals. The single RNN classifier (e.g., LSTM or GRU) seems to work better than its counterparts due to its excellent ability to model temporal dependencies. Yet here, the complex hybrid models indeed outperform the single component. For example, Golmohammadi *et al.* (2017c) achieves a better specification than (Schirrmeister *et al.*, 2017) on the same data set because of combing with RNN. Most of the epileptic seizure detection models claim a rather high classification accuracy (above 95%). One possible reason for this is their use of the binary classification scenario, which is much easier than multiclass classification.

There are only a small number of publications on the brain signal-controlled smart environment, and the brain signals used were collected using very diverse methods. This is growing, promising field because it is easy to integrate with smart home and smart hospital to benefit healthy persons and those with physical impairments alike. Another benefit of brain signal research is that it bridges people's inner and outer worlds through communication techniques. In this area, many investigations focus on the VEP signals because the visual evoked potential is obvious and easily detected; one important data source is from the third BCI competition. In addition, brain signal analysis would represent a significant advance in security systems since the brain signals are invisible and very difficult to mimic. Their resistance to spoofing makes them a rising star in the area of identification/authentication in scenarios in which confidentiality is a concern. The main drawbacks of brain signal-based security systems are that the equipment is both expensive and inconvenient (e.g., the subject must wear an EEG headset to permit monitoring of brain waves).

Affective computing has drawn much attention in recent years. EEG signals have high temporal resolution and are able to capture rapidly varying emotions. Therefore, almost all these studies focus on spontaneous EEG signals recorded while the subject is watching video intended to arouse a specific emotion. In some cases, the EOG signals are also captured for eye-blinking artifact removal. Another reason for this interest is that there are several open-source EEG-based affective analysis data sets available (e.g., DEAP and SEED). EEG-based affective computing contains two main streams. One focuses on developing powerful discriminative classifiers (such as hierarchical CNN) capable of performing feature extraction and classification in the same step. The other trains deep representative models (e.g., DBN-RBM) to recognize latent features, and then sends the learned representations to a classical classifier (such as HMM or MLP). It can be observed that the former models (Hiroyasu *et al.*, 2014; Zhang *et al.*, 2018b) seem to outperform the latter (Zheng and Lu, 2015) by a small margin on the SEED data set.

Driver fatigue detection could be easily integrated into mobile platforms for self-driving vehicles, however, there are only a few publications in this area due to the high costs of research and the lack of an accessible data set. Many other interesting potential applications for deep learning models (e.g., the guilty knowledge test and gender detection) have also been explored.

PART 3

Recent Advances on Deep Learning for EEG-Based BCI

Chapter 6

Robust Brain Signal Representation Learning

6.1 Overview

Different subjects may produce different brain signal patterns, even in the same objective situation; this is known as "intersubject variability" (Saha and Baumert, 2019) and is caused by a number of person-specific factors (such as age, gender, and fatigue) whose fundamental relevances are not yet fully understood (Kanoga *et al.*, 2020). In order to build generalized BCI systems, it is nevertheless important to extract robust brain signal representations that can work across subjects (Atum *et al.*, 2019). BCI systems can be divided into three categories in terms of the subject-based generalization:

- **Subject-Dependent.** The training and the testing sets are collected from the *same subject*. In other words, the well-trained BCI system should work well on one specific subject but may not work at all on others.
- **Cross-Subject.** The training set and testing set are acquired from the *same group of subjects*. As a result, the BCI system can serve any subject from this group but its performance may drop dramatically when applied to individuals beyond the group.
- **Subject-Independent.** The training set is gathered from a number of subjects, whereas the testing set is collected from an unseen person who was not part of the training set. This type of BCI system is able to learn robust and generic representations from the training set and thus to build a predictor that should perform well on any unseen subject.

Obviously, a good BCI system should be able to handle the subject-independent scenarios so that it can be deployed in real-world applications. However, it is very difficult to build such a robust system due to inter-subject noise. As a result, researchers must add constraints and study within narrow parameters (e.g., subject-dependent and cross-subject) in order to shed light on subject-independent developments. This chapter introduces the details for each category of system and the corresponding typical deep learning technologies.

Brain Signal Decomposition. The collected brain signals can be divided into three components: the pattern-related component (e.g., the action/status of the subject), the component evoked by objective factors (e.g., environmental noise), and the component evoked by subjective factors (e.g., gender, height, age, health). Taking EEG signals as an example, the collected raw data E is formulated as:

$$E = E_p + E_o + E_s \tag{6.1}$$

where E_p, E_o, and E_s separately denote the aforementioned three components and in which E_p is the signal of interest in BCI system, E_o is the noise caused by intrasubject variability, and E_s is the noise caused by intersubject variability.

Taxonomy of Brain Signal Representations. Generally, the information contained in brain signals is threefold: temporal information along the time axis, spatial information along with the channel distribution, and graphical information indicated by the brain connectivity network. As described in Section 4.7, RNN and its variants are able to extract temporal features; CNN and its variants can explore the latent spatial features; and graph neural networks (GNNs) discover the graphical features based on the connections among brain regions.

The abovementioned three categories of deep neural networks can not only work separately but also integrate with each other. Separately, they function as three types of representation learning algorithms: RNNs, CNNs, and GNNs. By combining two of these, we can create: RNNs+CNNs to model spatiotemporal dependencies, RNNs+GNNs to consider both temporal and graphical information, and CNNs+GNNs to discover spatial features and brain region connections. Finally, it is also possible to combine all three, RNNs+CNNs+GNNs, to analyze the information as a whole.

When applying any of the aforementioned seven types of robust representation learning frameworks, it is worth noting that: (1) RNNs and CNNs have been widely used to handle brain signals since 2016, whereas GNNs only caught on in 2018; so there are only a few studies that adopt GNNs for brain signal analysis (which is why GNNs are not included in Chapter 3). The independent use of GNNs and the integration of GNNs with other deep learning models present a bright prospect for future research. (2) Furthermore, we cannot definitively say which method is most effective or whether combined-model performs better than single-model method. The performance of the algorithms highly depends on the signal type, scenario, resolution, and so on. But we have reason to believe that the brain signals with higher temporal resolution (e.g., EEG and fNIRS) can provide more information to temporal-analysis models while brain signals with a higher spatial resolution (e.g., fMRI) are better suited to CNNs. (3) When combining multiple deep learning components, the framework must be correctly designed to avoid conflicts between them. For example, if CNN would smash the graphical information, a combined CNN+GNN model might perform worse than the single CNN. (4) Apart from the basic models discussed above, the representative deep learning models (e.g., AE and RBM) can perform noise and dimensionality for any feature set, while the MLP structures can be used to learn representations and make predictions in any context. In this section, we focus on the basic deep learning frameworks; representative models and MLPs are covered further on.

Terminology. Before launching into the more detailed information, we here briefly introduce some terms commonly used to describe the brain signals. EEG recordings are long and continuous and cannot be directly fed into the downstream stages. Therefore, we slice these long recordings into short *segments,* each of which is called a *sample.* This operation is called *segmentation.* In machine learning, we investigate the characteristics of data at the segment level and attempt to find the characteristic feature of multiple samples in the same category. Thus, if we found that an unseen sample contained the specific feature, we could infer that this unseen sample belonged to a particular category.

There are two key parameters in segmentation: the *time window* and *the overlap.* The time window is the length of the segment, while the overlap indicates the interval during which two adjacent segments coincide. For example, assume we have an EEG recording lasting 10 s and that segmentation is conducted with a time window of 1 s and an overlap of 50%. Therefore, we can harvest $19 = (10/1) * (1/50\%) - 1$ segments, where each segment lasts 1 s. The first segment is sliced from 0 to 1s, the second segment is sliced from 0.5 s to 1.5 s, and so on. The smallest scale is the time point, also called the time slice, which refers to the signal collected signal in a single *sampling operation*. For instance, assume we have equipment with a sampling frequency of 100 Hz associated with 64 channels/electrodes. That means the equipment will take 100 samples per second and that each sampling will produce 64 values (one value per electrode). The vector which contains these 64 elements is called a *time point*. It clearly follows that each one-segment will contain 100 *time points* and this can be represented as a matrix with 100 rows and 64 columns, in which, each row corresponds to a time point and each column corresponds to an EEG channel. Taking this one step further, we can regard the horizontal (rows) as the temporal axis and the vertical (columns) as the spatial axis.

6.2 Subject-Dependent

Most of the existing BCI-related studies focus on subject-dependent scenarios, which allows researchers to bypass the challenge of intersubject variability to concentrate solely on representation learning. Relatively speaking, the subject-dependent BCI systems are, therefore, the primary systems.

As shown in Equation (6.1), in the subject-dependent situation, $\boldsymbol{E}_s$ approximates to zero, as all the signals are gathered from the same subject. The deep learning models only need to learn the latent features, from $\boldsymbol{E}$, that can represent the interested component (i.e., $\boldsymbol{E}_p$). Most methods presented in Table 4.1 are developed for representation learning and classification in subject-dependent BCI systems. Next, we will introduce several typical deep learning-based robust representation learning algorithms.

6.2.1 *Temporal Representation Learning*

It is essential to discover the latent temporal representations in order to describe the pattern of brain activities. Straightforwardly put, RNN, along with its variants, provides the opportunity to explore temporal representations that indicate how the brain signal varies over time.

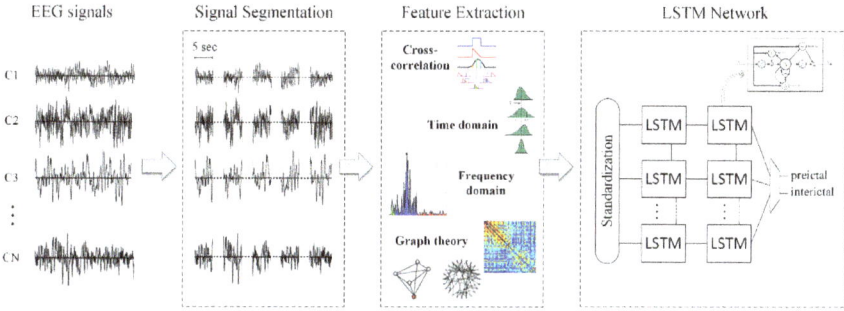

Fig. 6.1: Illustration of an epileptic seizure predicting system utilizing temporal representation learning (Tsiouris *et al.*, 2018). The details of LSTM cell are presented in Fig. 3.4(a).

Figure 6.1 presents a flowchart that combines traditional feature engineering with advanced deep recurrent neural networks to predict epileptic seizure (Tsiouris *et al.*, 2018). The proposed system contains four stages: EEG signal collection, signal segmentation, feature extraction, and LSTM-based prediction. In the first stage, the continuous EEG streams are gathered by specific equipment. Then the raw EEG signals are fed into the segmentation stage by setting the time window as 5 s, with an overlap of 0. The time window and overlap are empirically determined, depending on the experimental scenario, signal quality, and even equipment sampling frequency. Generally, a time window of 30 s is set for sleep EEG, 0.1–3 s for MI EEG, and 2–10 s for emotional and mental disease EEG. The overlap usually ranges from 0 to 50%. The outcome of the segmentation is that we have a number of segments in each channel (i.e., C1). Please note that there still exist temporal relationships among these segments since they are split along the time axis.

Next, the segments are fed into the third stage, which is feature engineering. This workflow shows how features are manually extracted, including the cross-correlation feature, time domain features, frequency domain features, and graph-theoretic measures. The cross-correlation feature measures the correlations among each pair of EEG channels to quantify the similarity between any two EEG signals, taking into consideration the potential time delays between two spatially distant signals. In the time domain, the commonly used features in brain signal analysis include several statistical moments (mean value, variance, skewness, kurtosis), the zero crossings, which show the number of sign changes; the peak-to-peak voltage, which indicates the difference between the highest/lowest amplitude in a segment; as well as the total signal area, which measures the absolute area under signal. In addition, the spectral information of the EEG signals is also taken into consideration, since various frequency domain features are also extracted, including the total energy spectrum and the energy percentage across the fundamental rhythmic bands (e.g., Delta, Theta, Alpha, Beta, and Gamma bands). Finally, this framework also considers the graph-theoretic measures, which reflect the functional connectivity between the EEG channels. However, the features extracted from the channel connections are pretty shallow, so that this work is *not* regarded as graphical information learning.

Afterward, the extracted features flow to the LSTM-powered classifier to predict the seizure states (i.e., preictal or interictal). Based on preliminary experiments, we have selected the LSTM classifier design as follows. The classifier is comprised of six layers: the first LSTM layer contains 643 cells, where each cell receives one feature extracted in the previous stage. Next, a dropout layer with a dropout rate of 0.5 is adopted to prevent overfitting. This dropout layer randomly sets 50% of the cell outputs as zero to force the LSTM to learn more generalized representations. The third layer is another LSTM layer containing 128 cells, followed by a dropout layer with a 0.2 dropout rate. The output of this second dropout layer is fed into a fully connected layer with 30 nodes, which is by an output layer with only two nodes. The output from this layer is the seizure state prediction. If the first element is larger, it means the input EEG segment belongs to the preictal phase; otherwise, the EEG segment belongs to the interictal phase.

Note that, the manual feature extraction in the third stage is essential in traditional BCI systems but can be omitted in deep learning-based BCIs because deep learning can automatically learn features from raw EEG signals. Tsiouris *et al.* (2018) compare the performance of the LSTM classifiers under two scenarios. On the first occasion, the LSTM classifier receives raw EEG signals and directly identifies the epileptic seizure stage without feature learning. On the second occasion, the LSTM classifier takes the manually extracted feature as its basis for prediction. The experimental results show that the second classifier outperforms the first with a margin of around 10%. This indicates that combining of traditional features with state-of-the-art deep learning models is a good way to boost the performance of BCI systems.

6.2.2   *Spatial Representation Learning*

Due to the blooming of research on computer vision, massive models are developed to extract spatial features from images. Lots of them, such as ResNet (Pereira *et al.*, 2018) and SincNet (Zeng *et al.*, 2019; Ravanelli and Bengio, 2018), can be adopted and modified to learn spatial representations of brain signals. In this section, we present several typical spatial feature learning frameworks which already used in BCI area.

Let us continue take epileptic seizure detection as the application scenario, CNN has been adopted to capture spatial features from both the time-domain and frequency-domain of EEG signals. As shown in Figure 6.2, the workflow illustrates how CNN captures the latent representations from raw EEG data and use them to predict the seizure state (Zhou *et al.*, 2018).

The spatial representation learning framework contains five stages: input the EEG data in both the time and frequency domains, the convolutional operation, mean pooling, the fully connected, and the final output layers. It can be clearly seen that this system requires two sets of inputs, one from the time domain and the other from the frequency

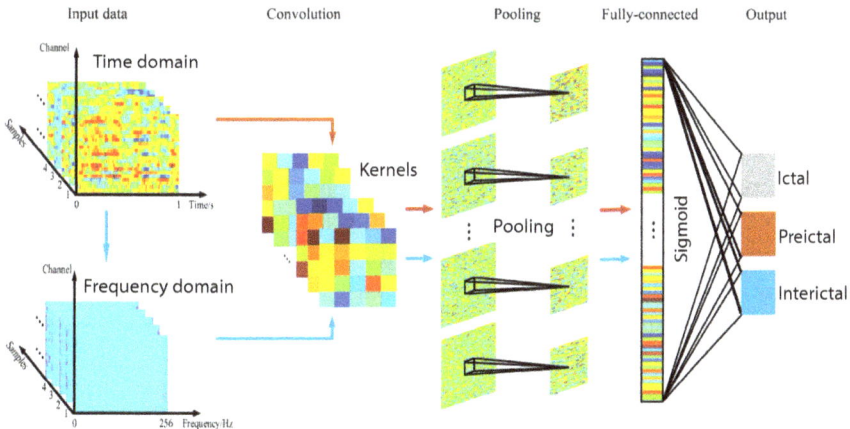

Fig. 6.2: Framework of a CNN-based model for brain signal processing (Zhou *et al.*, 2018).

domain. The frequency domain segments are transformed from the time domain through fast Fourier transform (FFT). The input data are drawn as 3D, with a time/frequency dimension, a channel dimension, and a sample dimension, whereas in fact, they prove to contain only two dimensions if we take a closer look at a single sample.

The convolutional, pooling, and fully connected layers are introduced in Section 3.1.3; let's very briefly recap their key features. The CNN structure primarily includes convolutional layers, pooling layers, and fully connected layers. Convolutional layers apply a convolution operation to the input, transferring the result to the next layer. The convolution emulates the response of an individual neuron to visual stimuli. Convolutional networks may include local or global pooling layers that combine the outputs of neuron clusters from one layer into a single neuron in the next layer. Mean pooling uses the average value from each cluster of neurons in the previous layer. Fully connected layers connect every neuron in one layer to every neuron in the next layer. The last fully connected layer indicates the prediction results and is called the output layer.

Figure 6.2 presents a multichannel time series based on either time or frequency domain signals directly input into a CNN as the input layer. Structurally, CNNs are made up of convolutional layers interspersed with pooling layers, followed by fully connected layers. The CNN model we use consists of three main layers: the convolutional layer, which has six feature maps connected to the input layer via 5×5 kernels and consists of kernels that slide across the EEG signals; the second layer, which comprises a 2×2 mean pooling layer and is mainly used to extract key features and to reduce the computational complexity of the network; and the final fully connected layer, which outputs the classification result (i.e., ictal, preictal, or interictal) using sigmoid activation (Zhou *et al.*, 2018).

Compared to a conventional CNN, the highlight of this framework is its two branches of input. Specifically, not only the usual spatial information in time domain is considered; representations in the frequency domain are also captured. Although frequency-domain features are always taken into account in traditional BCI systems, they have been considered in only a few deep learning-based systems.

In the schematic shown in Figure 6.2, the features from the time and frequency domains are integrated into the convolutional layer through concatenation. Beyond that, there are two further options for integrating information from multiple sources, giving us, in total, three potential methods of integration: early, middle, and late integration (Zitnik *et al.*, 2019).

In early integration, information is combined in the input layer. For example, we have a sample with shape $[100, 64]$ in the time domain and, by performing FFT, obtain a shape $[100, 64]$ in the frequency-domain. Next,they are fused into a matrix with shape $[100, 128]$ and then use CNN is used to analyze the newly formed matrix. The middle integration method, as illustrated in Figure 6.2, combines features in hidden layers. In the late integration method, analysis is independently performed on each input data and the model results are combined to make the final prediction. For example, we could build two identical CNN classifiers, where one receives time domain samples and the other receives frequency domain samples, and then, the average of the output layer of the two classifiers is taken as the final predicted output.

EEGNET (Lawhern *et al.*, 2018) is a popular network that can explicitly learn spatial relationships from raw EEG signals. The model contains several branches to learn different aspects of the input EEG signals in parallel and at last combine the learned representation for final classification. In detail, there is a trainable kernel in each branch transforming the raw EEG data to a new space corresponding to a specific EEG frequency band. This strategy is similar to the filter bank in traditional EEG feature extraction. Afterwards, a depthwise convolution layer is used to learn spatial filter. A separable convolution is then designed to decouple the relationship across feature maps by optimally merging the outputs.

6.2.3 *Graphical Representation Learning*

Deep learning models generally treat the EEG channels as a 2D grid input, neglecting the complex connections among the EEG channels (Li *et al.*, 2019). By contrast, the graphical information refers to the information carried by the connections among brain regions. The graphical representation analysis is based on graph theory, according to which each brain region (e.g., EEG electrode) is represented as a vertex, while the connection among regions is represented by an edge. Figure 6.3 presents the fully-connected graph (i.e., complete graph) in which each vertex is connected with all the other vertices. Figure 6.3 (left) presents the graphical connection among five channels in the frontal (F) and temporal (T) lobes of the human brain. The exact connections among brain regions are still not definitely understood by biological and neurological researchers; thus, the connectivities might, within reason, be different. There were some studies that tried to learn a graph structure (i.e., the connectivities among

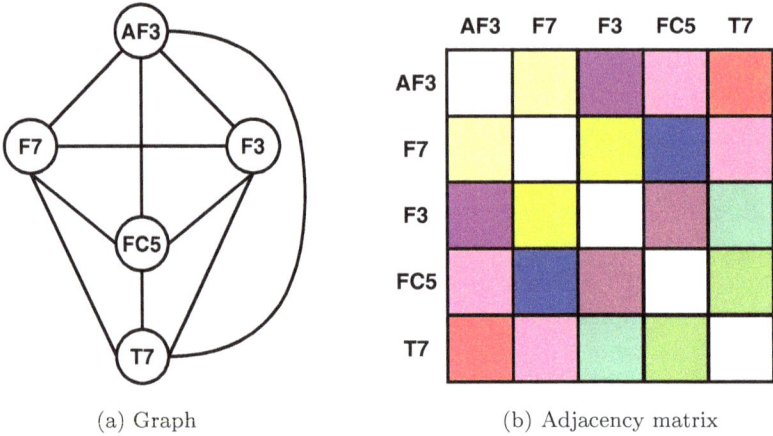

(a) Graph   (b) Adjacency matrix

Fig. 6.3: Example of a complete weighted undirected graph with five vertices and its corresponding adjacency matrix. The five vertices are reading from the Frontal (F) and temporal (T) lobes of the human brain. The adjacency matrix is a symmetric matrix in which the colors denote the connection weights.

vertices); Li *et al.* in particular proposed an iterative method based on data aggregation in a graph convolution mechanism to reveal the latent graph structure (Li *et al.*, 2019).

The right side of Figure 6.3 provides the adjacency matrix, where each element indicates the connection relationship among the channel pairs. In mathematical terms, assume $\boldsymbol{A}$ denotes the adjacency matrix and the element $\boldsymbol{A}_{F7\cdot T7}$ represents the connection state between channel $F7$ and channel $T7$. The adjacency matrix could be binary or numeric. In a binary adjacency matrix, each element is either a 1 or a 0. Let's say $\boldsymbol{A}_{F7\cdot T7} \in \{0,1\}$, where 1 indicates that channel $F7$ and channel $T7$ are connected and 0 indicates that they are not. If the adjacency matrix is numeric, i.e., $\boldsymbol{A} \in \mathcal{R}^{N \times N}$ where N denotes the number of channels, the value of $\boldsymbol{A}_{F7\cdot T7}$ indicates the edge weight. The adjacency matrix shown in Figure 6.3 is numeric.

GNNs are a subset of deep learning models that are designed to deal with graphical data. GNNs have drawn much attention in the past two years, and there are already numerous variants; please refer to Wu *et al.* (2020) for more detailed models. Here, we briefly introduce a typical GNN model called the Graph Convolutional Network (GCN) (Kipf and Welling, 2017). Compared to a standard CNN, the key modification of a GCN is a novel developed convolutional operation called the graph convolutional operation. Rather than using kernels, the graph convolutional operation aggregates information from the neighborhood. More specifically, the graph convolution consists of three steps. First, the "information" in each channel is calculated through a nonlinear transformation. This step is formulated as

$$h_{F7} = \sigma(W E_{F7} + b) \qquad (6.2)$$

where E_{F7} denotes the input of the $F7$ channel. Assuming a segment with a one-second time window and a sampling frequency of 100, then E_{F7} is a real-valued vector with 100 elements, where each element is sampled at $F7$ channel at a time point. The h_{F7} denotes the "information" carried by channel $F7$. The σ, W, and b denote nonlinear activation function, trainable weights, and biases, respectively. The second step is aggregation of the information from the neighbors of $F7$. Since this is a fully connected graph, the neighboring nodes contain $AF3, F3, FC5$, and $T7$. The aggregated information h'_{F7} is calculated by

$$h'_{F7} = \text{AGG}(w_{AF3}h_{AF3}, w_{F3}h_{F3}, w_{FC5}h_{FC5}, w_{T7}h_{T7}) \qquad (6.3)$$

where AGG is the aggregation function, such as sum, average, and maximum and w_{AF3} represents the importance weights (or edge weight) of channels $AF3$ to $F7$. The weight is calculated as

$$w_{AF3} = \frac{1}{\sqrt{D_{AF3}}\sqrt{D_{F7}}} \qquad (6.4)$$

where D_{AF3} denotes the "degree" of vertex $AF3$ meaning the amount of $AF3$'s neighbors. The last step is "update," calculated as

$$E'_{F7} = \text{UPD}(h_{F7}, h'_{F7}) \qquad (6.5)$$

where UPD is a an update function generally, the sum function. The above three steps compose one GCN layer. A GCN model contains an arbitrary number of GCN layers (in practice, usually two).

Next, let's inspect a real framework in Figure 6.4 (Song *et al.*, 2018). The input graph denotes the connectivities among EEG channels. Here, the adjacency matrix is built based on predefined brain connectivity, as shown in Figure 6.5. Each frequency band can form a new graph signal based on the input graph. All the graphs share the same structure but have different features. A GCN layer extracts graphical representations from each input graph, which are then combined together by middle integration. The final predictions can be obtained by application of the ReLU activation, fully connected, and softmax layers.

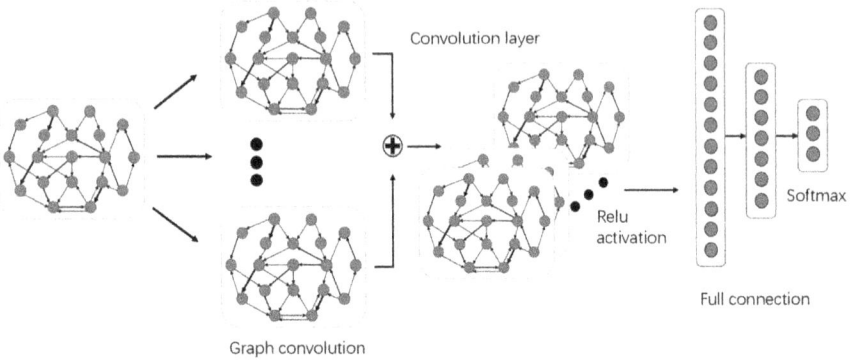

Fig. 6.4: The framework of the GCN model for EEG representation learning (Song *et al.*, 2018).

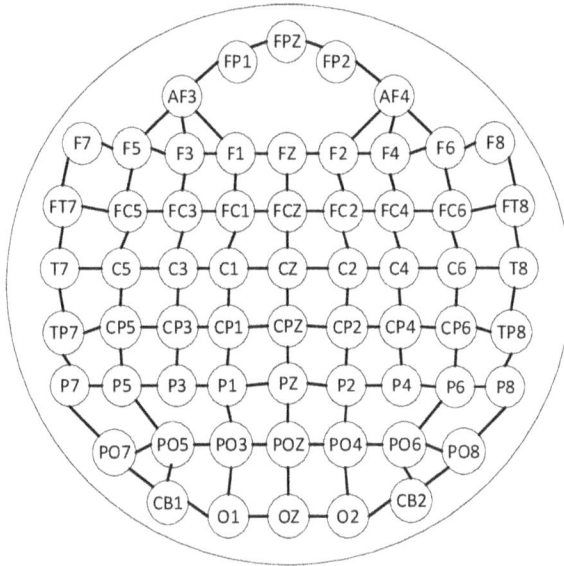

Fig. 6.5: Illustration of the connections among EEG channels. The adjacency matrix is built based on this connection map (Song *et al.*, 2018).

For increased understanding, we present a more detailed pipeline that shows how the GNN may be used as the core algorithm to build a practical system. As shown in Figure 6.6, Li and Duncan (2020) propose a GNN-based framework for mapping regional and cross-regional functional activation patterns for classification tasks, such as distinguishing patients

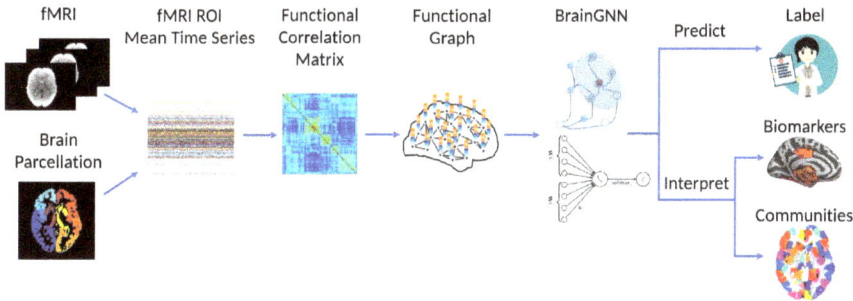

Fig. 6.6: The pipeline of GNN-based fMRI analysis (Li and Duncan, 2020).

with neurological disorders from healthy control subjects and performing cognitive task decoding. The first stage, as in Figure 6.6, is to build the brain graph from the input fMRI data. The brain regions of interest (ROIs) are regarded as vertices, while the functional connectivity, between the ROIs, computed as the pairwise correlations of fMRI time series, are regarded as edges. The second stage is to exploit graphical information obtainedfrom the functional graph. The BrainGNN is a GNN variant modified to aggregate node information only from selected neighbors rather than the entire neighborhood. After that, the BrainGNN can predict the task label (e.g., patient or healthy control).

6.2.4 *Spatiotemporal Representation Learning*

It is intuitive to combine RNNs and CNNs to capture the spatiotemporal representations of brain signals since both models have achieved great success in this area. A large number of publications exploit spatiotemporal features to analyze brain signals in multiple applications such as person identification (Wilaiprasitporn *et al.*, 2019) and motor imagery recognition (Zhang *et al.*, 2018a).

To prevent conflict between temporal and spatial features within a single EEG sample, researchers have developed an effective method for converting 1D EEG time points to a 2D matrix. As shown in Figure 6.7, a 32-channel cap can collect a vector with 32 elements at each time point. The elements are distributed following the 10–20 system (left) to form a matrix of 9 rows and 9 columns (right). The gray circle on the right side is set as 0, whereas the brown circles inherit the values of collected elements. In this way, the 1D EEG time point with shape $[1, 32]$ is converted to a matrix with

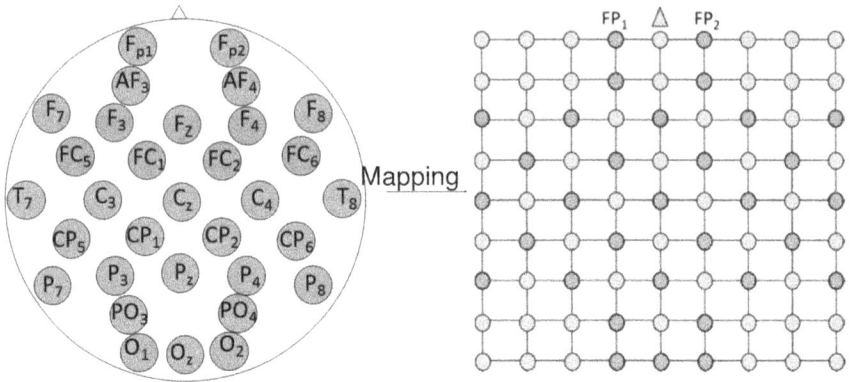

Fig. 6.7:   Convert 1D EEG time point to a 2D matrix (Wilaiprasitporn *et al.*, 2019).

shape $[9, 9]$. Assuming an EEG segment contains 10 time points (with shape $[10, 32]$), the above transformation maps the 2D segment to 3D format with shape $[10, 9, 9]$. Next, we show how the transformation aids spatiotemporal representation learning in specific cases.

We here introduce two typical frameworks combining RNN and CNNs in cascade and parallel, respectively. The former learns temporal and spatial representations sequentially, while the latter learns temporal and spatial representations in parallel and combines them for prediction.

The cascade R-CNN model, Figure 6.8, is composed of four stages: the data input, spatial representation learning, temporal representation learning, and calculation of the prediction results. The model receives a single EEG sample (with shape $[10, 9, 9]$) containing both spatial and temporal information and makes a five-class classification that associates the input sample with a specific motor-imagery action (closing both eyes, moving both feet or both fists, or moving left or right fist). In the first stage, the EEG sample is split into 10 time points where each time point is a matrix with shape $[9, 9]$. In the second stage, a standard CNN structure is adopted to learn the latent spatial correlations from each EEG sample. In this example, 10 2D CNNs are adopted, and each CNN receives one time point and extracts a spatial feature vector. Note that all the CNN feature extractors share the same parameters. In the stage which follows, the learned spatial representations are fed into a temporal feature extractor composed of two LSTM layers. The LSTM layers receive spatial information that are acquired in different time points, and produce temporal-spatial representations. Lastly, the spatiotemporal representation

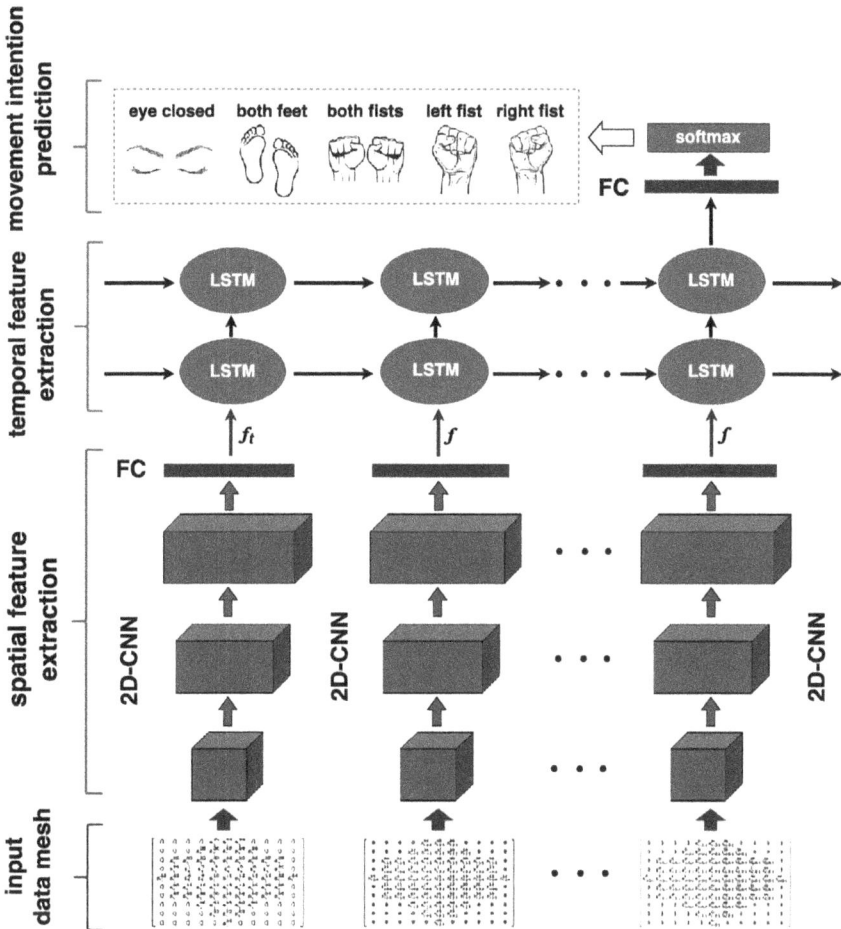

Fig. 6.8: The framework of spatiotemporal representation learning of EEG signals in cascade (Zhang *et al.*, 2018a).

flows into a fully connected layer and is then passed through a softmax layer to make the prediction.

On the other hand, the parallelized R-CNN in Figure 6.9 is composed of the input data, the spatial and temporal and spatial feature extraction in parallel, and the future fusion followed by intent prediction stages. In contrast to the cascade model, this framework extracts the spatial and temporal representations simultaneously and performs a middle-integration to fuse features in order to make the final prediction.

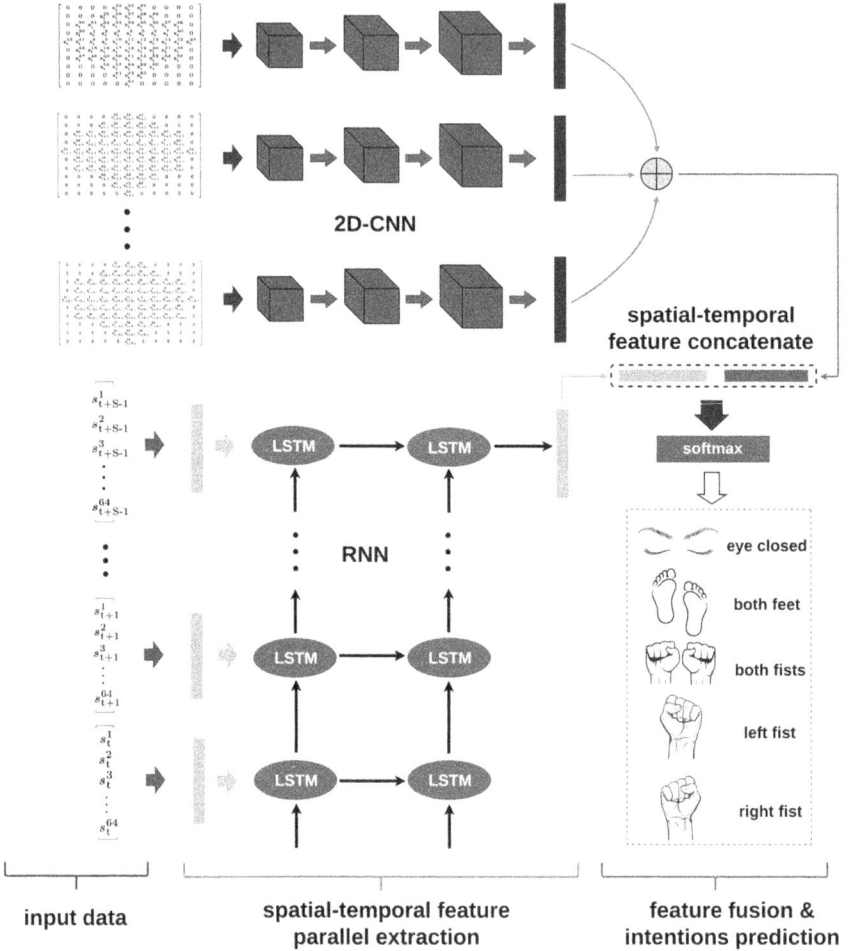

Fig. 6.9: The framework of spatiotemporal representation learning of EEG signals in parallel (Zhang *et al.*, 2018a).

6.2.5 *Discussion*

In principle, in addition to the four methods introduced above, there are three further potential approaches to representation learning: temporal-graphical, spatial-graphical, and spatiotemporal graphical. We list them here for completeness, although many blanks that need filling in this area. (This is also why we did not include GNNs in Chapter 3.) We have ourselves made a preliminary exploration of these topics. In this book, we propose a BCI system which captures spatial graphical representations to print a replica of an imagined object (see Section 10.1).

6.3 Cross-Subject

6.3.1 *Overview*

This section describes how to learn representations in the cross-subject context. Basically, cross-subject BCI systems are pretty similar to subject-dependent; the former simply expand the scope from one specific subject to a group of subjects. However, recalling Equation (6.1), under the cross-subject situation, the model must also handle the E_s caused by inter-subject variability.

In principle, the deep learning models introduced in the previous section could also be used to perform cross-subject tasks, but they would not perform well due to intersubject variability. There are two main strategies which address this issue. The first is fairly straightforward: enhancing the data size so as to force the machine learning model to learn the latent patterns. This strategy is not only costly in terms of data, as it requires massive brain signal samples, it cannot even theoretically guarantee a good prediction accuracy. The second strategy is to design a deep learning framework capable of extracting purified representations (from which the noise caused by intersubject variability has been removed).

Next, we introduce a deep learning framework for effectively learning EEG features across multiple subjects–meaning that training and testing samples have all been randomly selected from a group of subjects. In particular, we present a novel intent recognition approach to classifying cross-subject and multiclass EEG data. First we analyze the similarity of EEG signals and calculates the correlation coefficients matrix for both interclass and cross-subject conditions. Next, we extract EEG signal features by applying the autoencoder algorithm. The features are then fed into a XGBoost classifier which identifies the EEG data categories, where each category corresponds to a specific intent.

6.3.2 *EEG Characteristic Analysis*

To increase our knowledge of EEG data characteristics in preparation for further EEG classification, we quantify the similarity between EEG samples by calculating their Pearson correlation coefficients, following:

$$\rho(A, B) = \frac{1}{N-1} \sum_{i=1}^{N} \left(\frac{A_i - \mu_A}{\sigma_A} \right) \left(\frac{B_i - \mu_B}{\sigma_B} \right), \quad i = 1, 2, \ldots, N \quad (6.6)$$

where A and B denote two EEG samples, each containing N elements; μ_A and σ_A denote the mean and standard deviation of A,respectively and μ_B

and σ_B denote the mean and standard deviation of B, respectively. The Pearson correlation coefficient is positively correlates with similarity, and both are in the range of $[0, 1]$.

We here introduce two similarity concepts used in our measurement: self-similarity and cross-similarity. Self-similarity is defined by the similarity of EEG signals within the same EEG category, whereas cross-similarity is defined as the similarity of EEG signals from two different EEG categories. Both self-similarity and cross-similarity are measured under two conditions: interclass and cross-subject.

Interclass Measurement. Under the interclass situation, we measure the correlation coefficient matrix for *each specific subject* and calculate the average matrix by calculating the mean value of the entire matrix. For example, there are five classes for a specific subject, we calculate a $5 * 5$ correlation coefficient matrix. In this matrix, $\rho_{\breve{i},\breve{j}}$ denotes the correlation coefficient between the samples from the class $\breve{i}$ and the samples from the class $\breve{j}$. Self-similarity indicates the similarity between two different samples from the same class. Cross-similarity indicates the average similarity of each possible *class pair* of samples belonging to one *specific subject*.

Cross-Subject Measurement. Under the cross-subject situation, we measure the correlation coefficients matrix for *each specific* class and then calculate the average matrix. Self-similarity indicates the similarity between two different samples from the same class of the same subject. Cross-similarity indicates the average similarity of each possible *subject pair* of samples belonging to the *specific class*.

Table 6.1 shows the interclass correlation coefficient matrix and the corresponding statistical self- and cross-similarity. The last column (PD)

Table 6.1: Interclass correlation coefficients matrix. The correlation coefficients matrix (upper left section) is the average of 20 correlation coefficients obtained from 20 unique subjects. Self-S, Cross-S, and PD denote self-similarity, cross-similarity, and percentage difference, respectively.

Class	0	1	2	3	4	Self-S	Cross-S	PD
0	0.4010	0.2855	0.4146	0.4787	0.3700	0.401	0.3872	3.44%
1	0.2855	0.5100	0.0689	0.0162	0.0546	0.51	0.1063	79.16%
2	0.4146	0.0689	0.4126	0.2632	0.3950	0.4126	0.2854	30.83%
3	0.4787	0.0162	0.2632	0.3062	0.2247	0.3062	0.2457	19.76%
4	0.3700	0.0546	0.3950	0.2247	0.3395	0.3395	0.3156	7.04%
Range	0.1932	0.4938	0.3458	0.4625	0.3404	0.2038	0.2809	75.72%
Average	0.3900	0.1870	0.3109	0.2578	0.2768	0.3939	0.2680	28.05%
STD	0.0631	0.1869	0.1334	0.1487	0.1255	0.0700	0.0932	27.33%

provides the percentage difference between the self-similarity and cross-similarity. We can observe from the results that the self-similarity is always higher than the cross-similarity for all classes, meaning that the samples' intraclass cohesion is stronger than their interclass cohesion. The percentage difference shows a noticeable fluctuation, indicating the varying intraclass cohesion over different class pairs.

Similarly, Table 6.2 provides the cross-subject correlation coefficient matrix and gives an alternative visualization of the results. Again, we find that, for each class, the self-similarity is higher than cross-similarity, with varying percentage differences. The standard deviations of cross-similarity for each of the five classes are similar. This indicates a steady and even distribution of the data set among subjects and classes.

The above analysis results satisfy our following hypothesis for cross-subject classification: (1) Self-similarity is consistently higher than cross-similarity under both interclass and cross-subject conditions; (2) the higher the interclass percentage difference, the better the classification results; and (3) lower average percentage differences and standard deviations of the subjects result in better classification performance under the cross-subject condition.

6.3.3 *Representation Learning Framework*

In this section, we review the proposed intent recognition algorithm by first normalizing the input EEG data and then automatically exploring the feature representation of the normalized data. Then, we adopt an XGBoost classifier to classify the trained features. The methodology flowchart is shown in Figure 6.10.

6.3.3.1 *Normalization*

Normalization plays a crucial role in a knowledge discovery process when dealing with parameters of different units and scales. For instance, assuming that one input feature ranges from 0 to 1, whereas another ranges from 0 to 100, without adjustment, the analysis results would be dominated by the latter feature. Generally, there are three widely used normalization methods: min–max normalization, unity normalization, and z-score scaling (also called standardization).

Min–Max Normalization. Min–max normalization projects all of the elements in a vector into the range of $[0, 1]$. This method maps features

Table 6.2: Cross-subject correlation coefficients matrix. STD denotes Standard Deviation, SS denotes Self-similarity, CS denotes Cross-similarity, and PD denotes Percentage Difference.

subjects	Class 0			Class 1			Class 2			Class 3			Class 4		
	SS	CS	PD	SS	CS	PD	SS	CS	PD	SS	CS	PD	SS	CS	PD
subject1	0.451	0.3934	12.77%	0.2936	0.1998	31.95%	0.3962	0.3449	12.95%	0.4023	0.1911	52.50%	0.5986	0.4375	26.91%
subject2	0.3596	0.2064	42.60%	0.3591	0.1876	47.76%	0.5936	0.3927	33.84%	0.2354	0.2324	1.27%	0.3265	0.2225	31.85%
subject3	0.51	0.3464	32.08%	0.3695	0.2949	20.19%	0.3979	0.3418	14.10%	0.4226	0.3702	12.40%	0.4931	0.4635	6.00%
subject4	0.3196	0.1781	44.27%	0.4022	0.1604	60.12%	0.3362	0.2682	20.23%	0.4639	0.3905	15.82%	0.3695	0.2401	35.02%
subject5	0.4127	0.2588	37.29%	0.3961	0.2904	26.69%	0.3128	0.2393	23.50%	0.4256	0.1889	55.62%	0.3958	0.3797	4.07%
subject6	0.33	0.2924	11.39%	0.3869	0.3196	17.39%	0.3369	0.3281	2.61%	0.4523	0.1905	57.88%	0.4526	0.3321	26.62%
subject7	0.4142	0.3613	12.77%	0.3559	0.342	3.91%	0.3959	0.3867	2.32%	0.4032	0.3874	3.92%	0.4862	0.2723	43.99%
subject8	0.362	0.1784	50.72%	0.4281	0.2121	50.46%	0.4126	0.2368	42.61%	0.3523	0.1658	52.94%	0.4953	0.2438	50.78%
subject9	0.324	0.2568	20.74%	0.3462	0.2987	13.72%	0.3399	0.3079	9.41%	0.3516	0.1984	43.57%	0.3986	0.177	55.59%
subject10	0.335	0.1889	43.61%	0.3654	0.2089	42.83%	0.2654	0.2158	18.69%	0.3326	0.2102	36.80%	0.3395	0.2921	13.96%
subject11	0.403	0.1969	51.14%	0.3326	0.2066	37.88%	0.3561	0.3173	10.90%	0.4133	0.1697	58.94%	0.5054	0.44	12.94%
subject12	0.4596	0.2893	37.05%	0.4966	0.3702	25.45%	0.3326	0.2506	24.65%	0.4836	0.3545	26.70%	0.3968	0.3142	20.82%
subject13	0.3956	0.2581	34.76%	0.4061	0.3795	6.55%	0.3965	0.3588	9.51%	0.3326	0.1776	46.60%	0.3598	0.3035	15.65%
subject14	0.3001	0.299	0.37%	0.3164	0.2374	24.97%	0.4269	0.3763	11.85%	0.3856	0.1731	55.11%	0.4629	0.3281	29.12%
subject15	0.3629	0.3423	5.68%	0.3901	0.2278	41.60%	0.7203	0.2428	66.29%	0.3623	0.3274	9.63%	0.3862	0.3303	14.47%
subject16	0.3042	0.1403	53.88%	0.3901	0.3595	7.84%	0.4236	0.331	21.86%	0.4203	0.1634	61.12%	0.4206	0.3137	25.42%
subject17	0.396	0.1761	55.53%	0.3001	0.2232	25.62%	0.6235	0.3579	42.60%	0.5109	0.198	61.24%	0.3339	0.2608	21.89%
subject18	0.4253	0.3194	24.90%	0.3645	0.2286	37.28%	0.6825	0.222	67.47%	0.4236	0.3886	8.26%	0.4936	0.3017	38.88%
subject19	0.5431	0.3059	43.68%	0.3526	0.2547	27.77%	0.4326	0.3394	21.54%	0.5632	0.3729	33.79%	0.4625	0.219	52.65%
subject20	0.3964	0.3459	12.74%	0.3265	0.2849	12.74%	0.4025	0.3938	2.16%	0.3265	0.1873	42.63%	0.3976	0.2338	41.20%
Min	0.3001	0.1403	0.37%	0.2936	0.1604	3.91%	0.2654	0.2158	2.16%	0.2354	0.1634	1.27%	0.3265	0.177	4.07%
Max	0.5431	0.3934	55.53%	0.4966	0.3795	60.12%	0.7203	0.3938	67.47%	0.5632	0.3905	61.24%	0.5986	0.4635	55.59%
Range	0.2430	0.2531	55.16%	0.2030	0.2191	56.21%	0.4549	0.1780	65.31%	0.3278	0.2271	59.97%	0.2721	0.2865	51.53%
Average	0.3902	0.2667	31.40%	0.3689	0.2643	28.14%	0.4292	0.3126	22.96%	0.4032	0.2519	36.84%	0.4288	0.3053	28.39%
STD	0.0644	0.0723	0.1695	0.0456	0.0636	0.1518	0.1223	0.0589	0.1853	0.0717	0.0890	0.2066	0.0690	0.0759	0.1485

EEG Data   **Feature transformation**

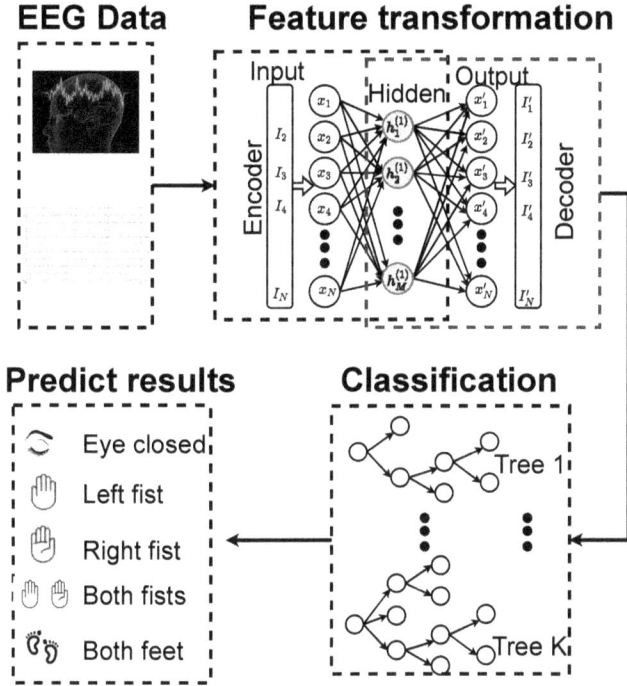

Fig. 6.10: The methodology flowchart. The collected EEG data flow into the Feature Representation component which seeks the appropriate representation and interpretation. The I_i and I'_i respectively indicate the input and output EEG data. The X_i, h_i, and x'_i indicate the neurons in the input layer, the hidden layer, and the output layer, respectively. The learned feature representation h will be sent to an XGBoost classifier with K trees. The prediction result produced by the classifier corresponds to the user's intent, indicating the user's movement intention (to close their eyes or move their left or right hand, etc.).

to the same range, disregarding of their original means and standard deviations. The formula for min–max normalization is given below:

$$x_{new} = \frac{x - x_{min}}{x_{max} - x_{min}} \tag{6.7}$$

where x_{min} and x_{max} respectively denote the minimum and maximum of feature x.

Unity Normalization. Unity normalization rescales the features by the percentage or the weight of each single element. It calculates the sum of all the elements and then divides each element by that sum. The normalization equation is as follows:

$$x_{new} = \frac{x}{\sum x}$$ (6.8)

where $\sum x$ denotes the sum of feature x. Similar to min–max normalization, the results of this method also belong to the range of $[0, 1]$.

Z-score Scaling. *Z*-score scaling forces features into normal Gaussian distribution (zero mean and unit variance):

$$x_{new} = \frac{x - \mu}{\sigma}$$ (6.9)

where μ and σ denote the expected and standard deviation of the input feature x, respectively.

The experimental results — for complete details refer to Zhang *et al.* (2017c) — show that the *z*-score scaling normalization obtains the best classification performance, whereas the unity normalization obtains the worst. Without going into detailed explanation, all of the remaining experiments in this book use the *z*-score scaling method.

6.3.3.2   *Feature Representation*

To exploit the deep correlation between EEG signals, we adopt an autoencoder in order to learn a more accurate representation of EEG. The autoencoder (Nguyen *et al.*, 2015) is an unsupervised machine learning algorithm that aims to explore a lower-dimensional representation of high-dimensional input data for dimensionality reduction. Structurally, the autoencoder is a multilayer backpropagation neural network that contains three types of layers: the input layer, the hidden layer, and the output layer. The layer which processes the input data into a compressed representation in the hidden layer is called the encoder while the procedure that reconstructs the compressed representation from the hidden layer to the output layer is called the decoder. Both the encoder and the decoder yield a set of weights W and biases b. An autoencoder is called a basic autoencoder (when there is only one hidden layer) or a stacked autoencoder (when there are multiple hidden layers). In our prior experience, a basic autoencoder works better than a stacked autoencoder when dealing with EEG signals; therefore, we adopt the basic autoencoder structure here.

Let $\mathcal{X} = \{X_i | i = 1, 2, \ldots, N\}, X_i \in \mathbb{R}^d$ be the entire training data (unlabeled), where X_i denotes the i-th sample, N denotes the number of training samples, and d denotes the number of elements in each sample.

The $h_i = \{h_{ij}|j = 1, 2, \ldots, M\}, h_i \in \mathbb{R}^M$ represents the learned feature in the hidden layer for the ith sample, where M denotes the number of neural units in the current layer (the number of elements in h_i). For simplicity's sake, we use x and h to represent the input data and the data in the hidden layer, respectively. First, the encoder transforms the input data x to the corresponding representation h by the encoder weights W_{en} and the encoder biases b_{en}:

$$h = W_{en}x + b_{en} \tag{6.10}$$

Then, the decoder transforms the hidden layer data h to the output layer data x' by the decoder weights W_{de} and the decoder biases b_{de}:

$$x' = W_{de}h + b_{de} \tag{6.11}$$

The function of the decoder is to reconstruct the encoded feature h and make the reconstructed data x' as similar to the input data x as possible. The discrepancy between x and x' is calculated by the MSE (mean squared error) cost function which is optimized by the RMSPropOptimizer.

In summary, the process of training an autoencoder is the task of optimizing the parameters to achieve the minimum cost between the input x and the reconstructed data x'. The hidden layer data h contains the refined information. Such information can be regarded as representation of the input data, which is also the final outcome of autoencoder. In above formulation, the dimension of the input data x and the refined feature (the hidden layer data h) are d and M, respectively. The autoencoder entails either dimensionality reduction if $d > M$ or dimensionality increase if $d < M$.

6.3.3.3 *Intent Recognition*

To recognize the intent reflected by the represented feature, in this section, we employ the XGBoost classifier (Chen and Guestrin, 2016) classifier. XGBoost, or Extreme Gradient Boosting, is a supervised scalable tree boosting algorithm derived from the concept of the gradient boosting machine (Friedman, 2001). Compared with the gradient boosting algorithm, XGBoost proposes a more regularized model formalization to prevent overfitting, with the engineering goal of pushing the limits of computational resources for boosted tree algorithms to achieve better performance. Consider n sample pairs $D = \{(x_{i'}, y_{i'})\}$, $(|D| = n, x_{i'} \in \mathbb{R}^m, y_{i'} \in \mathbb{R})$ where $x_{i'}$ denotes a m-dimensional sample and $y_{i'}$ denotes the corresponding label. XGBoost aims to predict the label $\tilde{y}_{i'}$ of every given sample $x_{i'}$.

The XGBoost model is an ensemble of a set of classification and regression trees (CART), each having its leaves and corresponding scores. The final results of such a tree ensemble is the sum of all its individual trees. For a tree ensemble model of K' trees, the predicted output is:

$$\tilde{y}_{i'} = \sum_{k'=1}^{K} f_{k'}(x_{i'}), \quad f_{k'} \in F \tag{6.12}$$

where F is the space of all trees, and $f_{k'}$ denotes a single tree.

The objective function of XGBoost includes loss function and regularization. The loss function evaluates the difference between each ground truth label $y_{i'}$ and the prediction result $\tilde{y}_{i'}$. It can be chosen based on various conditions such as cross-entropy, logistic regression, and mean square error. The regularization function is the most outstanding contribution of XGBoost. It calculates the complexity of the model and applies a larger penalty to a more complex structure.

The objective function is defined as:

$$\Psi = \sum_{i}^{n} l(\tilde{y}_{i'}, y_{i'}) + \sum_{k'}^{K} \Omega(f_{k'}) \tag{6.13}$$

where $l(\tilde{y}_{i'}, y_{i'})$ is the loss function and $\sum_{k'} \Omega(f_{k'})$ is the regularization term. The complexity of a single tree is calculated as

$$\Omega(f_{k'}) = \gamma T + \frac{1}{2}\lambda\|\omega\|^2 \tag{6.14}$$

where T is the number of leaves in the tree, $\|\omega\|^2$ denotes the square of the L2-norm of the weights in the tree, γ and λ are the coefficients. The regularized objective function helps to deliver both a simple structure model and predictive functions. More specifically, the first term, Ω, penalizes complex structures of the tree (fewer leaves result in a smaller Ω), while the second term penalizes the overweighting of individual trees to prevent the overbalanced trees dominating the model. Moreover, the second term helps to smooth the learned weights and avoid overfitting.

6.4 Subject-Independent

The subject-independent scenario in the design of BCI systems has proven exceedingly challenging and is thus a topic still at the frontiers of BCI research. This issue, to our best knowledge, has yet to attract sufficient attention due to its high degree of difficulty. Thus, up to now, there are not many effective models for solving this problem, but the emerging transfer learning methods are beginning to shed some light. Next, we will briefly introduce transfer learning and its application in the BCI system.

6.4.1 *Transfer Learning*

Transfer learning is a subdomain of machine learning that takes knowledge learned from one task (i.e., the source task) and reuses it in performing another similar task (i.e., the target task) (West *et al.*, 2007). In general, the transfer learning procedure contains two stages. In the first stage, a deep learning model (i.e., source model) is trained on a large and informative data set. In the second stage, the pretrained model is slightly modified,then assigned a similar task on a smaller data set. The idea behind transfer learning is that we expect the pretrained model to have learned some general rules in one specific area (e.g., object recognition) which it can then use to boost performance in a related scenario.

For instance, assume researchers train an object recognition model (e.g., CNN) on a massive data set with millions of images belonging to a thousand categories, deliberately excluding cat. Then, we make a slight modification to the pretrained CNN–here, changing the output layer to obtain a new CNN model. They next feed images of cats and dogs to the new CNN model with the aim of teaching it to recognize a cat. In this example, the source task is object recognition of a thousand objects, the learned general knowledge is ability to differentiate among objects, and the target task is to recognize a cat.

6.4.2 *Intersubject Transfer Learning*

Intuitively, transfer learning can be adopted to solve the problem of subject-independent classification. That is, we can design a source deep learning model trained on a variety of subjects and then tune the model to work on an unseen subject. In this way, the general knowledge about the task will be transferred to the new subject. It worth noting that, in contrast to the standard transfer learning model, the intersubject transfer learning model is not assigned a new task. The source model and the tuned model work on the same classification task, but they are tested over different samples.

There are only a few studies which use intersubject transfer learning techniques to handle BCI systems (Fahimi *et al.*, 2019; Tan *et al.*, 2018;

Völker *et al.*, 2018). Here, we introduce a representative transfer learning framework proposed in Tu and Sun (2012). In this approach a new model is trained on the data from a pool of subjects and then transfers the learned knowledge to a new subject. In other words, it aims to use data from prior subjects (i.e., source subjects) to reduce the training burden on the current user (i.e., target subject).

The key to this framework is extracting and transferring knowledge from the source subject to the target subject. To begin with, a candidate filter bank is built based on the Extreme Energy Ratio (EER) criterion (Sun, 2008). The EER learns spatial filtering that maximizes the variance of spatially filtered EEG samples belonging to one class while minimizing the variance of other classes.

Given the filter bank, the second step is to learn two subsets of filters: the robust filter bank and the adaptive filter. The foundation of these so-called filters is a linear transformation of the input sample based on the EER algorithm. Each filter can recover a source signal from the source samples, but different signals produced by different filters have different significance. Some signals are not sensitive to subject, that is, the signals don't contain $\boldsymbol{E}_s$. These signals can be used to learn generic features related only to brain activities. The corresponding filters are selected as a robust filter bank. Other signals may be adaptive to the target subject, their the corresponding filters can be regarded as an adaptive filter bank. (Refer to Tu and Sun (2012) for further details on how to select for the robust and adaptive filter banks.)

Furthermore, the EEG data set from each subject is mapped onto two new data sets, where one is transformed by the robust filter bank and another is transformed by the adaptive filter bank. Then, for each subject, there are two classifiers, a robust classifier and an adaptive classifier, each separately trained on one of the two new data sets. To this end, for K source subjects, we have K robust classifier and K adaptive classifier. Take the robust-classifier as an example, a dynamic ensemble strategy is employed to combine the outcomes of the K classifiers. The ensemble strategy can assign different weights to the outcomes of different classifiers during combination. The weights are calculated by measuring the similarity of the target subject and the source subject whose data were used to train the classifier. Finally, the two classifiers are incorporated into a complementary ensemble. A learned parameter is used to trade off the robustness and adaptiveness in the final prediction.

Chapter 7

Cross-Scenario Classification

7.1 Overview

BCI systems can be widely applied to a range of scenarios, from mental disease diagnosis to emotion recognition to movement intention prediction. However, most of the existing studies are scenario-specific. This means that the designed BCI system — in particular, the core classification algorithm — can only work on a specific scenario and fails at other tasks. For example, a classification model can attain great performance in neurological disorder diagnosis but fail at emotion recognition. Consequently, it is necessary to develop algorithms that can learn the most informative representations relevant to each the specific scenario.

Cross-scenario classification requires the algorithm to automatically capture the distinctive feature (i.e., learn the discriminative pattern) from the input data regardless of scenario. In this chapter, the meaning of "cross-scenario" is twofold. First, the algorithm should achieve competitive performance across signal sources. For instance, the model is fed a combination of EEG and fNIRS signals in one scenario but a combination of EEG and smartphone signals in another. Second, the algorithm should work well across applications. In other words, the model is designed to handle the task of disease diagnosis when trained on a neurological data set; likewise, it can predict a user's emotion when trained on an affective data set.

In this chapter, we present two steam to handle cross-scenario classification: the attention mechanism and transfer learning. (Please refer to Section 6.4.1 for the background of transfer learning.) Here, we briefly introduce the basics of the attention mechanism.

Attention Mechanisms. Use of an attention mechanism enables an algorithm to pay varying amounts of attention to (i.e., weight differently) different components of the input data (Vaswani *et al.*, 2017). For instance, the different channels in EEG signals make unequal contributions to the prediction task. By adopting an attention mechanism, the algorithm can boost classification accuracy by assigning higher weights to the relevant channels and lower weights to the irrelevant ones.

The attention mechanism was first proposed in natural language processing and is now widely applied to many research topics. The key to adopting an attention mechanism is knowing how to calculate the attention weights/coefficients. The attention weights can either be automatically learned through backpropagation or linearly measured based on similarity within the input sample. Please refer to Vaswani *et al.* (2017) for more details.

In terms of BCI systems, when directed by an attention mechanism, the deep learning model can focus on the important channels or regions of interest in representation learning. As a result, the model will have high adaptability across different scenarios. For example, the frontal lobe is responsible for motor functions; as a result, the corresponding channels (e.g., F1 and F7) are of greater importance in predicting motor imagery intentions. By contrast, the occipital lobe is more sensitive to some visual stimuli, so the electrodes located at this lobe (e.g., O1 and O2) should be weighted when analyzing SSVEP signals. A well-designed deep learning framework combined with an attention mechanism can handle different tasks as long as the model is properly trained on the corresponding data sets.

7.2 Attention-Based Classification Across Signal Sources

7.2.1 *Overview*

In some cases, the performance of a BCI system will be improved by combining multiple brain signals (e.g., EEG, fNIRS, and EMG) and even other sensor signals (such as signals indicating the information of the user's activity, emotion, and location). Nowadays, diverse categories of sensors can be found in various wearable devices (most brain signals can be regarded as specific sensory signals). Such devices are now being widely applied in multiple fields, such as the Internet of Things (Kamburugamuve *et al.*, 2015; Zhang *et al.*, 2018h). In this section, we present a framework that takes advantage of both deep learning and reinforcement learning to learn informative representations across signal modality.

Compared to images and videos, sensory data are naturally formed as 1D signals, with each element accordingly representing a sensor channel. Several challenges face such sensor data classification. First, most existing classification methods use domain-specific knowledge and thus may become ineffective or even fail in complex situations where multimodal data are being collected (Bigdely-Shamlo *et al.*, 2015). For example, one approach that works well on fNIRS signals may not capable of dealing with EEG signals. Therefore, an effective and universal sensor data classification method is highly desirable for such complex situations. Any such framework would be expected to demonstrate both efficiency and robustness over a range of sensor signals.

Second, the wearable sensor data carries far less information than texts and images. For example, a sample signal gathered by a 64-channel EEG device contains only 64 numerical elements; hence, a more effective classifier would be required to extract discriminative information from such limited raw data. In the meantime, maximizing the utilization of the scarce available data demands cautious preprocessing and a rich fund of domain knowledge.

Inspired by the attention mechanism (Cavanagh *et al.*, 1992), we propose to concentrate on an attention zone of the signal to automatically learn the informative attention patterns for different sensor combinations. Here, the attention zone is a selection block of the signal of a certain length, sliding over the feature dimensions. Note that reinforcement learning has been shown to be capable of learning human-control on a variety of tasks (Mnih *et al.*, 2015). Reinforcement learning is employed to discover the attention zone. Considering that signals in different categories may have different interdimensional dependency (Markham and Townshend, 1981), we propose using the long short-term memory (LSTM) model (Long Short-Term Memory (Gers *et al.*, 1999; Zhang *et al.*, 2018a)) to exploit the latent correlation between signal dimensions. We propose a weighted average spatial LSTM (WAS-LSTM) classifier for exploring the dependencies in sensor data.

7.2.2 *Reinforced Selective Attention Model*

Suppose the input sensor data can be denoted by $\mathbf{X} = \{(\mathbf{x}_i, y_i),\ i = 1, 2, \ldots, I\}$ where $(\mathbf{x}_i, y_i)$ denotes the 1D sensor signal, called one *sample*, and I denotes the number of samples. In each sample, the feature $\mathbf{x}_i \in \mathbb{R}^K$ contains K elements and the corresponding ground truth y_i is an integer denoting the sample's category. $\mathbf{x}_i$ can be described as a vector with K elements, $\mathbf{x}_i = \{x_{ik}, k = 1, 2, \ldots, K\}$.

The proposed algorithm is shown in Figure 7.1. The main focus of the algorithm is to exploit the latent dependency between different signal dimensions. To this end, the proposed approach contains several components: (1) replication and shuffling; (2) selective attention learning; and (3) sequential LSTM-based classification.

7.2.2.1 *Motivation*

Learning to exploit the latent relationship between sensor signal dimensions is the main focus of the proposed approach. The signals belonging to different categories are assumed to have different interdimension dependent relationships, which contain rich and discriminative information. This information is critical to improving the discovery of distinctive signal patterns.

In practice, the sensor signal is often arranged as a 1D vector, which is less informative for the limited and fixed arrangement of elements. The order and the number of elements in each signal vector can affect their dependency. In many real-world scenarios, the multimodal sensor data are associated with the practical placement. For example, the EEG data are concatenated following the distribution of biomedical EEG channels. Unfortunately, the practical sensor sequence, with its fixed order and number, may not be suitable for interdimensional dependency analysis. Meanwhile, the optimal dimension sequence (Tan *et al.*, 2015b) varies with the sensor types and combinations. Therefore, we propose the following three techniques to address these drawbacks.

First, we replicate and shuffle the input sensor signal vector dimensionwise in order to provide as much latent dependency as possible among feature dimensions (Section 7.2.2.2).

Second, we introduce an attention zone as a selective attention mechanism, where the optimal interdimensional dependency for each sample depends on only a small subset of features. Here, the attention zone is optimized by deep reinforcement learning, which has proved to be stable and to perform well in policy learning (Section 7.2.2.3).

Third, we propose the WAS-LSTM classifier for extracting the distinctive interdimensional dependency (Section 7.2.2.4).

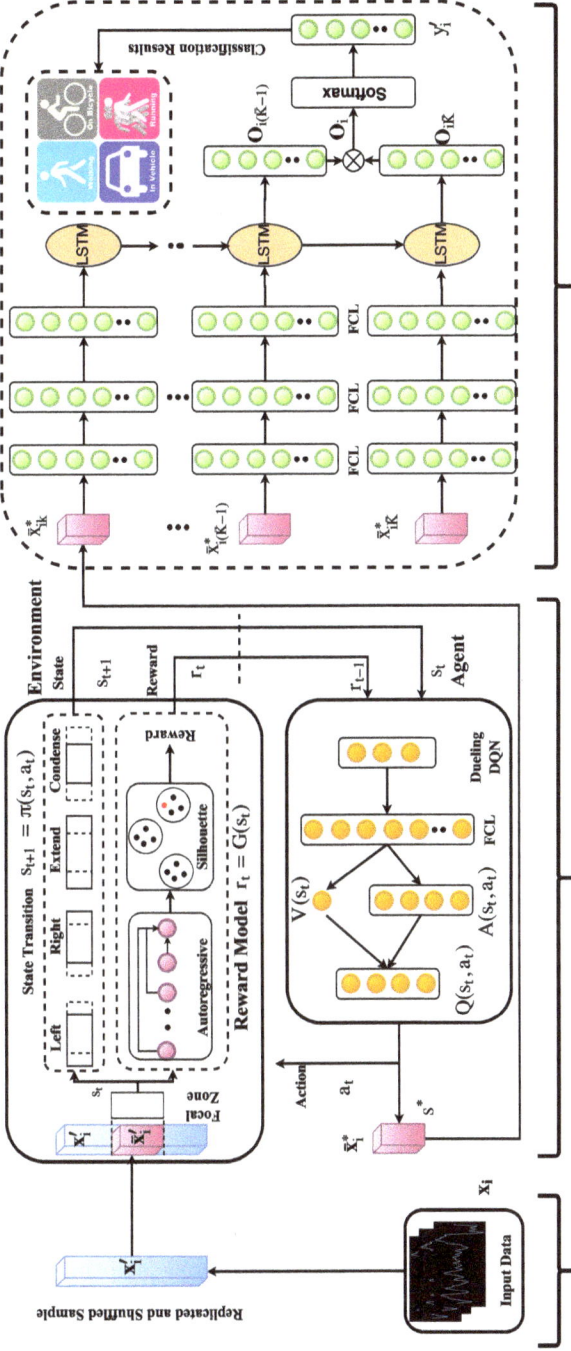

Fig. 7.1: Cross-modality representation learning workflow. The attention zone $\bar{\mathbf{x}}_i$ is a selected fragment of $\mathbf{x}_i'$ to be fed into the state transition and the reward model. In each step t, one action is selected by the state transition to update s_t based on the agent's feedback. The reward model evaluates the quality of the attention zone to the reward r_t. The dueling DQN is employed to find the optimal attention zone $\bar{\mathbf{x}}_i^*$ which will be fed into the LSTM-based classifier to explore the interdimensional dependency and predict the sample's label y_i'. FCL denotes fully connected layer. The state transition contains four actions: shift left, shift right, extend, and condense.

7.2.2.2   *Data Replication and Shuffling*

To provide as much information as possible, we have designed an approach to exploiting the spatial relationships among EEG signals. The signals belonging to different brain activities are assumed to have different spatially dependent relationships. We replicate and shuffle the input EEG signals dimensionwise. Using this method, all of the possible dimensional arrangements are equiprobable.

Suppose the input raw EEG data are denoted by $\mathbf{X} = \{(\mathbf{x}_i, y_i), i = 1, 2, \ldots, I\}$, where $(\mathbf{x}_i, y_i)$ denotes a single EEG sample and I denotes the number of samples. In each sample, the feature $\mathbf{x}_i = \{x_{ik}, k = 1, 2, \ldots, K\}, \mathbf{x}_i \in \mathbb{R}^K$ contains K elements corresponding to K EEG channels and $y_i \in \mathbb{R}$ denotes the corresponding label. x_{ik} denotes the kth dimension value in the ith sample.

In real-world collection scenarios, the EEG data are generally concatenated following the distribution of biomedical EEG channels. However, the biomedical dimension order may not present the best spatial dependency. The exhaustive method is too computationally expensive to produce all possible dimension arrangements. For example, a 64-channel EEG sample has $A_{64}^{64} = 1.28 \times 10^{89}$ combinations, which is an astronomical figure.

To provide more potential dimension combinations, we propose a method called replication and shuffling (RS). RS is a two-step mapping method which maps $\mathbf{x}_i$ to a higher dimensional space $\mathbf{x}_i'$ with complete element combinations: $\mathbf{x}_i \in \mathbb{R}^K \to \mathbf{x}_i' \in \mathbb{R}^{K'}, K' > K$.

In the first step (replication), $\mathbf{x}_i$ is replicated $h = K'/K + 1$ times. Then we get a new vector with a length of $h * K$ which is not less than K'; in the second step (shuffling), we randomly shuffle the vector replicated in the first step and intercept the first K' elements to generate $\mathbf{x}_i'$. Compared with $\mathbf{x}_i$, $\mathbf{x}_i'$ should theoretically contain more diverse dimension combinations. This RS operation should only be performed once for a specific input data set in order to provide a stable environment for the reinforcement learning to follow.

7.2.2.3   *Selective Attention Mechanism*

Inspired by the fact that the optimal spatial relationship depends only on a subset of feature dimensions, we introduce an attention zone that will focus on a fragment of those dimensions. Here, the attention zone is optimized by deep reinforcement learning, which has proved to be stable and to perform well in policy learning.

In particular, we aim to determine the optimal dimension combination, which includes the most distinctive spatial dependency among EEG signals. Since K', the length of $\mathbf{x}'_i$, is too large and computationally expensive, we introduce the attention mechanism (Cavanagh *et al.*, 1992) to balance the length and the information content, the effectiveness of which has already been demonstrated in recent research areas such as speech recognition (Chorowski *et al.*, 2015). We attempt to emphasize the informative fragment in $\mathbf{x}'_i$ and let $\bar{\mathbf{x}}_i$ denote the fragment, which is called *attention zone*. Let $\bar{\mathbf{x}}_i \in \mathbb{R}^{\bar{K}}$ and $\bar{K}$ denote the length of the attention zone which is automatically learned by the proposed algorithm. We employ deep reinforcement learning to discover the optimal attention zone (Mnih *et al.*, 2015).

As shown in Figure 7.1, the detection of the optimal attention zone includes two key components: the environment (including state transition and reward models) and the agent. Three elements (the state s, the action a, and the reward r) are exchanged in the interaction between the environment and the agent. All three elements are customized based on context in this study. Next, we introduce in detail the crucial components of our deep reinforcement learning structure:

- The **state** $\mathcal{S} = \{s_t, t = 0, 1, \ldots, T\}, s_t \in \mathbb{R}^2$ describes the position of the attention zone, where t denotes the time stamp. Since the attention zone is a shifting fragment on 1-D $\mathbf{x}'_i$, we design two parameters to define the state: $s_t = \{start^t_{idx}, end^t_{idx}\}$, where $start^t_{idx}$ and end^t_{idx} denote the start index and the end index of the attention zone,[1] respectively. In the training, s_0 is initialized as

$$s_0 = [(K' - \bar{K})/2, (K' + \bar{K})/2] \tag{7.1}$$

- The **action** $\mathcal{A} = \{a_t, t = 0, 1, \ldots, T\} \in \mathbb{R}^4$ describes which actions the agent could choose through which to act on the environment. Here at time stamp t, the state transition chooses one action to implement following the agent's policy π:

$$s_{t+1} = \pi(s_t, a_t) \tag{7.2}$$

---

[1] For example, for a random $\mathbf{x}'_i = [3, 5, 8, 9, 2, 1, 6, 0]$, the state $\{start^t_{idx} = 2, end^t_{idx} = 5\}$ is sufficient to define the attention zone as $[8, 9, 2, 1]$.

(a) Left Shifting

(b) Right Shifting

(c) Extend Shifting

(d) Condense Shifting

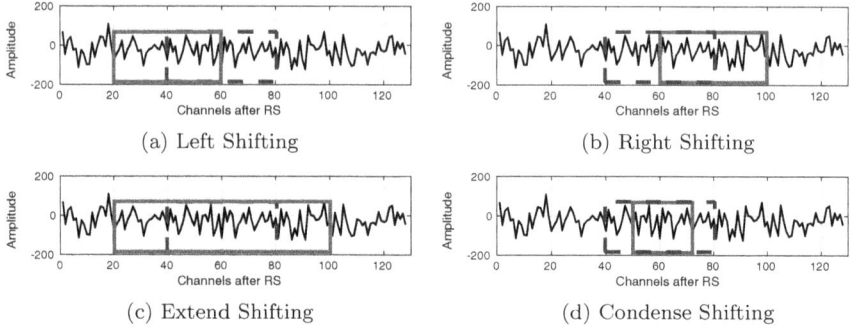

Fig. 7.2: The four actions in the state transition: shift left, shift right, extend, and condense. The dashed line indicates the attention zone before the action, whereas the solid line indicates the zone after the action.

In our case, we define four categories of actions (Figure 7.2) for the attention zone: shifting left, shifting right, extending, and condensing. For each action, the attention zone moves a random distance $d \in [1, d^u]$, where d^u is the upper boundary. For left shifting and right shifting actions, the attention zone shifts left or right with the step d; for the extending and condensing actions, both $start_{idx}^t$ and end_{idx}^t are moving d. Finally, if the state start index or end index lies beyond the boundary, a clip operation is conducted. For example, if $start_{idx}^t = -5$ which is lower than the lower boundary 0, we clip the start index as $start_{idx}^t = 0$.

- The **reward** $\mathcal{R} = \{r_t, t = 0, 1, \ldots, T\} \in \mathbb{R}$ is calculated by the reward model, which will be described later. The reward model Φ:

$$r_t = \Phi(s_t) \tag{7.3}$$

receives the current state and returns an evaluation as the reward.

Reward Model. Next, we introduce in detail the design of the reward model. The purpose of the reward model is to evaluate how the current state impacts the classification performance. Intuitively, the state which leads to better classification performance should have a higher reward: $r_t = \mathcal{F}(s_t)$. We set the reward modal $\mathcal{F}$ as a combination of the convolutional mapping and classification (Section 7.3.2.1). In the practical approach to optimization, the higher the classification accuracy, the more difficult it is to increase it yet further. For example, improving the accuracy at a higher level (e.g., from 90% to 100%) is much harder than at a lower level (e.g., from 50% to 60%). To encourage accuracy improvement at the higher level, we design a nonlinear reward function:

$$r_t = \frac{e^{acc}}{e-1} - \beta \frac{\bar{K}}{K'} \qquad (7.4)$$

where acc denotes the classification accuracy. The function contains two parts; the first part is a normalized exponential function with the exponent $acc \in [0,1]$, which encourages the reinforcement learning algorithm to search for the optimal s_t that leads to a higher acc. The motivation for the exponential function is that: *the reward growth rate* increases as the accuracy increases.[2] The second part is a penalty factor for the attention zone length to keep the bar shorter, and the β is the penalty coefficient.

In summary, the aim of the deep reinforcement learning is to learn the optimal attention zone $\bar{\mathbf{x}}_i^*$ which results in the maximum reward. The selective mechanism fully iterates $N = n_e * n_s$ times, where n_e and n_s denote the number of episodes and steps (Wang *et al.*, 2015b), respectively. ε-greedy method (Tokic, 2010) is employed in the state transition, which chooses a random action with probability $1 - \varepsilon$ or an action according to the optimal Q function $\text{argmax}_{a_t \in \mathcal{A}} Q(s_t, a_t)$ with probability ε. That is,

$$a_{t+1} = \begin{cases} \text{argmax}_{a_t \in \mathcal{A}} Q(s_t, a_t) & \varepsilon' < \varepsilon \\ \bar{a} \in \mathcal{A} & \text{otherwise} \end{cases} \qquad (7.5)$$

where $\varepsilon' \in [0,1]$ is randomly generated for each iteration, while $\bar{a}$ is randomly selected in $\mathcal{A}$.

For better convergence and quicker training, the ε is gradually increased with the iteration. The increment ε_0 follows:

$$\varepsilon_{t+1} = \varepsilon_t + \varepsilon_0 N \qquad (7.6)$$

Agent Policy and Optimization. The dueling deep Q networks (DQNs) (Wang *et al.*, 2015b) are employed as the optimization policy $\pi(s_t, a_t)$, which is enabled to efficiently learn the state-value function. The primary reason we employ a dueling DQN to find the best attention zone is that it updates all the four Q values at every step, whereas other policies only update one Q value per step. The Q function measures the expected sum of future rewards when taking that action and following the optimal policy

---

[2]For example, for the same accuracy increment of 10%, $acc : 90\% \to 100\%$ can earn a higher reward increment than $acc : 50\% \to 60\%$.

thereafter. In particular, for the specific step t, we have:

$$Q(s_t, a_t) = \mathbb{E}(r_{t+1} + \gamma r_{t+2} + \gamma^2 r_{t+3} \ldots)$$

$$= \sum_{n=0}^{\infty} \gamma^k r_{t+k+1} \tag{7.7}$$

where $\gamma \in [0,1]$ is the decay parameter that trades off between the importance of immediate and future rewards, while n denotes the subsequent number of steps. The value function $V(s_t)$ estimates the expected reward when the agent is in state s. The Q function is related to the pair (s_t, a_t) while the value function only associate with s_t.

Dueling DQN learns the Q function through the value function $V(s_t)$ and the advantage function $A(s_t, a_t)$, and then combines them as follows:

$$Q(s_t, a_t) = \theta V(s_t) + \theta' A(s_t, a_t) \tag{7.8}$$

where $\theta, \theta' \in \Theta$ are parameters in the dueling DQN network and are optimized automatically. Equation (7.8) is unidentifiable, which can be observed by the fact that we cannot recover $V(s_t)$ and $A(s_t, a_t)$ uniquely with the given $Q(s_t, a_t)$. To address this issue, we can force the advantage function to be equal to zero at the chosen action. That is, we let the network implement the forward mapping:

$$Q(s_t, a_t) = V(s_t) + [A(s_t, a_t) - \max_{a_{t+1} \in \mathcal{A}} (A(s_t, a_{t+1}))] \tag{7.9}$$

Therefore, for the specific action $a*$, if

$$argmax_{a_{t+1} \in \mathcal{A}} Q(s_t, a_{t+1}) = argmax_{a_{t+1} \in \mathcal{A}} A(s_t, a_{t+1}) \tag{7.10}$$

then we have

$$Q(s_{t+1}, a*) = V(s_t) \tag{7.11}$$

Thus, as shown in Figure 7.3 (the second last layer of the agent part), the stream $V(s_t)$ is forced to learn an estimation of the value function, while the other stream produces an estimation of the advantage function.

To assess the Q function, we optimize the following cost function at the ith iteration:

$$L_i(\Theta_i) = \mathbb{E}_{s_t, a_t, r_t, s_{t+1}}[(\bar{y}_i - Q(s_t, a_t))^2]$$

$$= \mathbb{E}_{s_t, a_t, r_t, s_{t+1}}[(\bar{y}_i - \theta V(s_t) + \theta' A(s_t, a_t))^2] \tag{7.12}$$

with

$$\bar{y}_i = r_t + \gamma \max_{a_{t+1}} Q(s_{t+1}, a_{t+1}) \tag{7.13}$$

The gradient update method is

$$\begin{aligned}
\nabla_{\Theta_i} L_i(\Theta_i) &= \mathbb{E}_{s_t, a_t, r_t, s_{t+1}}[(\bar{y}_i - Q(s_t, a_t))\nabla_{\Theta_i} Q(s_t, a_t)] \\
&= \mathbb{E}_{s_t, a_t, r_t, s_{t+1}}[(\bar{y}_i - \theta V(s_t) - \theta' A(s_t, a_t)) \tag{7.14} \\
&\quad \nabla_{\Theta_i}(\theta V(s_t) + \theta' A(s_t, a_t))]
\end{aligned}$$

7.2.2.4 *Weighted Average Spatial LSTM Classifier*

In this section, we propose WAS LSTM classification for two purposes. The first is to attempt to capture the cross-relationship among feature dimensions in the optimized attention zone $\bar{\mathbf{x}}_i^*$. The LSTM-based classifier is widely used for its excellent sequential information extraction ability, which is approved in several research areas such as natural language processing (Gers and Schmidhuber, 2001; Sundermeyer *et al.*, 2012). Compared to other commonly employed spatial feature extraction methods, such as CNNs, LSTM is less dependent upon the hyperparameter setting. However, the traditional LSTM focuses on the temporal dependency among a sequence of samples. Technically, the input data of traditional LSTM is 3D tensor shaped as $[n_b, n_t, \bar{K}]$ where n_b and n_s denote the batch size and the number of temporal samples, respectively. The WAS-LSTM aims to capture the dependency among various dimensions at one temporal point; therefore, we set $n_t = 1$ and transpose the input data as: $[n_b, n_t, \bar{K}] \rightarrow [n_b, \bar{K}, n_t]$.

The second advantage of WAS-LSTM is that it can be used to stabilize the performance of LSTM via the moving average method (Lipton *et al.*, 2015). Specifically, we calculate the LSTM outputs $\mathbf{O}_i$ by averaging the past two outputs rather than only the final one (Figure 7.1):

$$\mathbf{O}_i = (\mathbf{O}_{i(\bar{K}-1)} + \mathbf{O}_{i\bar{K}})/2 \tag{7.15}$$

The predicted label is calculated by $y_i' = \mathcal{L}(\bar{\mathbf{x}}_i^*)$ where $\mathcal{L}$ denotes the LSTM algorithm. The ℓ_2-norm (with parameter λ) is adopted as regularization to prevent overfitting. The sigmoid activation function is used on hidden layers. The loss function is cross-entropy and is optimized by the

---

Algorithm 1 Attention-Based Classification

Input: Sensor data $\mathbf{X}$
Output: Classification Results y_i'
  1: Initialization s_0
  2: **RS:** $\bar{\mathbf{x}}_i \leftarrow \mathbf{x}_i'$
  3: **Attention Patterns Learning:**
  4: **if** $t < N$ **then**
  5:    $a_t = Q(s_t, a_t)$
  6:    $s_{t+1} = \pi(s_t, a_t)$
  7:    $r_t = \mathcal{G}(s_t)$
  8:    $\bar{\mathbf{x}}_i^* \leftarrow \bar{\mathbf{x}}_i, a_t, s_t, r_t$
  9:    **return** $\bar{\mathbf{x}}_i^*$
10: **end if**
11: **LSTM-Based Classification:**
12: **if** $iteration < M$ **then**
13:    $y_i' \leftarrow \bar{\mathbf{x}}_i^*$
14:    **return** y_i'
15: **end if**
16: **return** y_i'

Adam optimizer algorithm (Kingma and Ba, 2014). The workflow of the attention-based multimodality signal classification is shown in Algorithm 1.

7.2.3 *Discussion*

First, the presented approach works better on high-dimensional feature space in that the information of interdimensional dependency is richer.

In addition, we introduce a novel idea that adopts an alternative reward model to estimate and replace the original reward model. In this way, the disadvantages of the original model, such as expensive computation, can be eliminated. The key is to keep the reward produced by the new model highly related to the original reward. The higher the correlation coefficient, the better. This sheds light on the possible combination of deep learning classifier and reinforcement learning.

Nevertheless, one weakness of this framework is that the reinforcement learning policy only works well in the specific environment in which the model is trained. The dimension indices should be consistent in the training and testing stages. Various policies should be trained according to different sensor combinations.

Furthermore, the proposed WAS-LSTM directly focuses on the dependency among the sensor dimensions and can produce a predicted label for each point, which provides the foundation for the quick-reaction online detection and other applications which require instantaneous detection. However, this model relies on a large enough number of signal dimensions to carry sufficient information to ensure accurate recognition.

7.3 Attention-Based Classification Across Applications

In the previous section, we proposed an attention-based learning framework which works well across signal sources. Next, we attempt to refine the task and focus on a general model capable of working with various types of brain signals, taking EEG as an example.

7.3.1 *Overview*

The accuracy and robustness of the EEG classification model enable the identification of cognitive activities in the realms of movement intention recognition, person identification, and neurological diagnosis with a high degree of confidence. Cognitive activity recognition systems (Vallabhaneni *et al.*, 2005) provide a bridge between the inner cognitive world and the outer physical world. Their use has recently become widespread in assisted living (Zhang *et al.*, 2018h), smart homes (Zhang *et al.*, 2017b), and the entertainment industry (Russoniello *et al.*, 2009); EEG-based person identification techniques empower the security systems implemented by banks or customs (Schetinin *et al.*, 2017; Zhang *et al.*, 2018g); EEG signal-based neurological diagnosis can be used to detect organic brain injury and abnormal synchronous neuronal activity such as epileptic seizure (Veeriah *et al.*, 2015; Acar *et al.*, 2007).

Apart from its widely known drawbacks such as time-consuming feature engineering and its dependency on domain knowledge, the classification of cognitive activity faces another significant challenge: The EEG signals have a low signal-to-noise ratio and are more chaotic than other sensor signals, such as those captured by wearable sensors. Thus, segment-based classification, which is widely used in sensor signal processing, may not suit cognitive activity recognition. A segment will contain continuous EEG samples clipped by the sliding window method (Fraschini *et al.*, 2015), whereas a single EEG sample (also called an EEG instance) is collected at a specific time point. Thus, segment-based classification has two drawbacks compared with sample-based classification. First, in a segment with many samples, the sample diversity may be offset by inverted samples, as EEG signals vary rapidly (Zhang *et al.*, 2019d). Second, segment-based classification requires more training data and a longer data-collection time. For example, suppose each segment has 10 samples without overlap; for the same training batch size, segment-based classification requires 10 times the data and the data-collection time as sample-based classification. As a result, segment-based classification cannot exploit the immediate intention of changing, and thus has low precision in practical deployment. For this reason, sample-based classification is more attractive.

To address the aforementioned issues, first, we propose a novel framework which can automatically learn distinctive features from raw EEG signals by developing a deep convolutional mapping component. Additionally, to grasp the characteristic information from different EEG application circumstances adaptively, we design a reinforced selective attention component that combines the benefits of an attention mechanism (Zhang *et al.*, 2018e) with those of deep reinforcement learning. Moreover, we sidestep the challenge posed by chaotic information by working on EEG samples instead of segments. The single EEG sample contains only spatial information without spatial cues.[3]

7.3.2 *Reinforced Attentive CNN*

We propose reinforced attentive CNN to directly classify raw EEG signals accurately and efficiently. As shown in Figure 7.3, the replicate and shuffle components and the reinforced selective attention component are very similar to Section 7.2.2.2 and Section 7.2.2.3. Here, we mainly introduce convolutional mapping.

7.3.2.1 *Convolutional Mapping*

For each attention zone, we further exploit the potential spatial dependency of selected features $\bar{\mathbf{x}}_i^*$. Since we focus on a single sample, the EEG sample

---

[3]We do not deny the usefulness of temporal information, but this section emphasizes spatial information, which may be easier to capture.

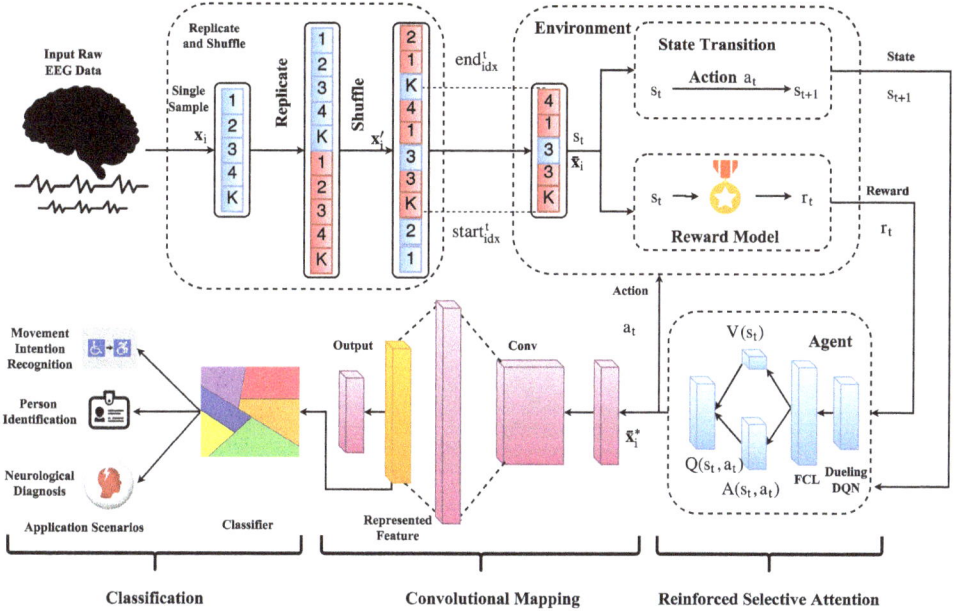

Fig. 7.3: Flowchart of the proposed approach. The input raw EEG sample $\mathbf{x}_i$ (K denotes the Kth element) is replicated and shuffled to provide more latent spatial combinations of feature dimensions. Then an attention zone $\bar{\mathbf{x}}_i$, which is a fragment in $\mathbf{x}'_i$, with the state $s_t = \{start^t_{idx}, end^t_{idx}\}$ is selected. The selected attention zone is input to the state transition and the reward model. In each step t, one action is selected by the state transition to update s_t based on the agent's feedback. The reward model evaluates the quality of the attention zone by the reward score r_t. The dueling DQN is employed to identify the best attention zone $\bar{\mathbf{x}}^*_i$ which will be fed into the convolutional mapping procedure to extract the spatial dependency representation. The represented features will be used for the classification. *FCL* denotes a fully connected layer. The reward model is the combination of the convolutional mapping and the classifier.

contains only a numerical vector with very limited information and is easily corrupted by noise. To address this drawback, we attempt to map the EEG single sample from the original space $\mathcal{O} \in R^K$ to a sparsity space $\mathcal{T} \in R^M$ by a CNN structure.

To extract as many potential spatial dependencies as possible, we employ a convolutional layer (Krizhevsky *et al.*, 2012) with many filters to scan the learned attention zone $\bar{\mathbf{x}}^*_i$. The convolutional mapping structure contains five layers (as shown in Figure 7.3): the input layer, which receives the learned attention zone, the convolutional layer, which is followed by one fully connected layer, and the output layer. The one-hot ground truth is compared with the output layer to calculate the training loss.

The ReLU nonlinear activation function is applied to the convolutional outputs. We describe the convolutional layer as follows:

$$x_{ij}^c = ReLU \left(\sum_{b=1}^{\bar{b}} W_c \bar{x}_{ij}^* \right) \quad (7.16)$$

where x_{ij}^c denotes the outcome of the convolutional layer, while $\bar{b}$ and W_c denote the length of filter and the filter weights, respectively. The pooling layer aims to reduce the redundant information in the convolutional outputs to decrease the computational cost. In our case, we try to preserve as much information as possible; therefore, our method does not employ a pooling layer. Then, in the fully connected layer and output layer

$$x_i^f = \text{ReLU}(W^f x_i^c + b^f) \quad (7.17)$$

$$y_i' = \text{softmax}(W^o x_i^f + b^o) \quad (7.18)$$

where W^f, W^o, b^f, b^o denote the corresponding weights and biases, respectively, y' denotes the predicted label. The cost function is measured by cross-entropy, and the ℓ_2-norm (with parameter λ) is adopted a regularization term to prevent overfitting:

$$cost = - \sum_x y_i' log(y_i) + \lambda \ell_2 \quad (7.19)$$

The Adam optimizer algorithm optimizes the cost function. The fully connected layer extracts as the represented features and feeds them into a lightweight nearest neighbors classifier. The convolutional mapping updates for N' iterations. The proposed cognitive activity recognition framework with reinforced attentive convolutional neural network is shown in Algorithm 2.

7.3.3 Evaluation Across Applications

Next, we evaluate the proposed method across three BCI applications.

7.3.3.1 Application Scenarios

Movement Intention Recognition (MIR). EEG signals measure human brain activities. Intuitively, different human intentions will produce

---

Algorithm 2 Cognitive Activity Recognition

Input: Raw EEG signals $\mathbf{X}$
Output: Predicted cognitive activity label y'_i
1: Initialization s_0
2: **RS:** $\bar{\mathbf{x}}_i \leftarrow \mathbf{x}'_i$
3: **Reinforced Selective Attention:**
4: **if** $t < N$ **then**
5:    $a_t = argmax_{a_t \in \mathcal{A}} Q(s_t, a_t)$
6:    $s_{t+1} = \pi(s_t, a_t)$
7:    $r_t = \mathcal{F}(s_t)$
8:    $\varepsilon_{t+1} = \varepsilon_t + \varepsilon_0 N$
9:    $\bar{\mathbf{x}}_i^* \leftarrow \bar{\mathbf{x}}_i, a_t, s_t, r_t$
10: **end if**
11: **Convolutional Mapping and Classifier:**
12: **if** *iteration* $< N'$ **then**
13:    $y'_i \leftarrow \bar{\mathbf{x}}_i^*$
14: **end if**
15: **return** y'_i

---

diverse EEG patterns (Zhang *et al.*, 2018h). Intention recognition plays a significant role in such practical scenarios as smart homes, assisted living (Zhang *et al.*, 2017b), brain typing (Zhang *et al.*, 2018h), and entertainment. For those with disabilities and the elderly, intent recognition can help them to interact with external smart devices such as wheelchairs or assistive robots via real-time BCI systems. For people incapable of speech, they may provide the chance to express their thoughts with the help of certain intention recognition technologies (e.g., brain typing). Even for healthy humans, intent recognition can be used in video game playing and other applications in daily life.

Person Identification (PI). EEG-based biometric identification (Schetinin *et al.*, 2017) is an emerging identification approach that is highly attack resistant. It has the unique advantage of avoiding or mitigating the threat of being deceived faced by other identification techniques. This technique can be deployed in identification and authentication scenarios such as bank security systems and customs security checks.

Neurological Diagnosis (ND). EEG signals collected from an individual in an unhealthy state differ significantly from those collected from an individual in the normal state as regards frequency and pattern of neuronal firing (Adeli *et al.*, 2007). EEG signals have therefore been used for neurological diagnosis for decades (Zhang *et al.*, 2019e). For example, epileptic seizure is a common brain disorder that affects around 1% of the population, and an EEG analysis of the patient allows detection of its octal state.

7.3.3.2 *Data Sets*

To evaluate how the proposed approach would perform in the afore-mentioned application scenarios, we chose several EEG data sets using various collection equipment, sampling rates, and data sources. We utilized motor imagery EEG signals from a public data set, the EEG motor movement/imagery database (*EEGMMIDB*), for intention recognition, the *EEG-S* data set for person identification, and the *TUH* data set for neurological diagnosis.

EEGMMIDB. We used the EEG data from PhysioNet EEGMMIDB, (*EEG motor movement/imagery database*) database, a widely used EEG database collected by the BCI 2000 instrumentation system (Goldberger *et al.*, 2000; Schalk *et al.*, 2004), to evaluate the proposed method. In particular, the data collected by the BCI 2000 system, which owns 64 channels and has an EEG data sampling rate of 160 Hz. During the collection of this database, the subject sits in front of one screen and performs the corresponding action as one target appears in different edges of the screen. According to the tasks, different annotations are labeled and can be downloaded from PhysioBank ATM. The actions in different tasks are:

Task 1: The subject closes his or her eyes and remains relaxed.

Task 2: A target appears at the left side of the screen, and the subject focuses on the *left hand* and imagines he/she is opening and closing his/her *left hand* until the target disappears.

Task 3: A target appears at the right side of the screen, and the subject focuses on the *right hand* and imagines he/she is opening and closing his/her *right hand* until the target disappears.

Task 4: A target appears at the top of the screen, and the subject focuses on *both hands* and imagines he/she is opening and closing both hands until the target disappears.

Task 5: A target appears at the bottom of the screen, and the subject focuses on *both feet* and imagines he/she is moving both feet until the target disappears.

Specifically, we selected 560,000 EEG samples from 20 subjects (28,000 samples from each subject) for our experiments. Each sample is one vector which includes 64 elements corresponding to 64 channels. Each sample corresponds to one intention (labeled from 1 to 5 are: eyes closed, left hand, right hand, both hands, and both feet).

EEG-S. EEG-S is a subset of EEGMMIDB, in which the data were gathered while the subject kept eyes closed and remained relaxed. Eight subjects were involved and each subject generated 7,000 samples. Labels are the subjects' IDs, ranging from 0 to 7.

TUH. TUH (Golmohammadi *et al.*, 2017a) is a neurological seizure data set of clinical EEG recordings associated with 22 channels from a 10/20 configuration. The sampling rate is set at 250 Hz. We selected 12,000 samples from each of 18 subjects. Half of the samples are labeled as epileptic seizure state (labeled as 1) and the remaining samples are labeled as normal state (labeled as 0). The experiment and parameter settings are the same as for the activity recognition applications.

7.3.3.3 *Evaluation Metrics*

For classification problems, there are a number of widely used criterion for evaluating the performance of the classification results, such as accuracy, precision, recall, F1 score, precision-recall curve (PRC), ROC (Receiver Operating Characteristic) curve, and AUC (Area Under the Curve). Before introducing the criterion, basic definitions related to classification problems include:

- true positive (TP): the ground truth is positive and the prediction is positive;
- false negative (FN): the ground truth is positive, but the prediction is negative;
- true negative (TN): the ground truth is negative and the prediction is negative;
- false positive (FP): the ground truth is negative, but the prediction is positive.

Based on these concepts, we define the criteria for evaluating the performance of the classification results as follows:

Accuracy: the proportion of all correctly predicted samples. Accuracy is a measure of how good a model is.

$$accuracy = \frac{TP + FN}{FP + FN + TP + TN} \tag{7.20}$$

The *test error* refers to the incorrectly predicted samples' proportion, which is equal to 1 minus accuracy.

Precision: the proportion of all positive predictions that are correctly predicted.

$$Precision = \frac{TP}{TP + FP} \tag{7.21}$$

Recall: the proportion of all real positive observations that are correctly predicted.

$$Recall = \frac{TP}{TP + FN} \tag{7.22}$$

F1 Score: a "weighted average" of precision and recall. The higher the F1 score, the better the classification performance.

$$F1\ Score = 2\frac{precision * recall}{precision + recall} \tag{7.23}$$

ROC: the ROC (Receiver Operating Characteristic) curve describes the relationship between TPR (True Positive Rate) and FPR (False Positive Rate) at various threshold settings. The TPR is generally called recall in machine learning. The FPR describes the proportion of all negative observations that are predicted incorrectly, which is a measure of the model's accuracy in predicting negative cases. One important advantage of ROC is that the curve is insensitive to the training data size. For example, the ROC curve of a classifier has no obvious fluctuation regardless of whether it is trained by 1 million samples or 10 billion samples.

One key concept behind the ROC curve is the predicted probability. The classifier produces the probability matrix in the prediction stage to illustrate the probability of the specific sample matching with the specific label. The ROC is plotted by a considerable number (FPR, TPR) of pairs which are created by traversal of the threshold through the whole probability range $[0, 1]$. The multiclass classification should transfer to binary classification (one-versus-all) by regarding one specific class as positive and all other classes as negative. Therefore, a multiclass classification task (e.g.,

K classes) will produce multiple ROC curves (K curves). In the diagonal of the ROC curve, where $TPR = FPR$, the classifier works as a random classifier and any sample has a 50% probability of being classified as positive or negative. The closer the ROC curve to the upper left corner, the better the performance of the ROC classifier.

AUC: the AUC (Area Under the Curve) represents the area under the ROC curve. The value of AUC drops in the range $[0.5, 1]$. On some occasions, the qualitative ROC curve is unable to clearly determine which classifier performs best. In these cases, the quantitative values AUC can more accurately evaluate classifier. The higher the AUC, the better the classifier.

7.3.3.4 *Overall Comparison*

To measure the accuracy of the proposed method, we compared against a set of baseline methods, including five nondeep learning and three deep learning baselines. Table 7.1 presents the classification metrics comparison between our approach and well-known baselines (including non-DL and DL baselines), where DL, AdaB, and LDA represent deep learning, adaptive boosting, and linear discriminant analysis, respectively. The results show that our approach achieved the highest accuracy on all the data sets. Specifically, the proposed approach achieved the accuracies of 0.9632, 0.9984, and 0.9975 on the EEGMMIDB, EEG-S, and TUH data sets, respectively. Furthermore, we conducted an ablation study by comparing our method, which mainly combined selective attention mechanism and a CNN, with the solo CNN. It turned out that our approach outperformed CNN, demonstrating that the proposed selective attention mechanism improved the distinctive feature learning. The confusion matrix and ROC curves (including the AUC scores) of each data set are given in Figure 7.4. In Figure 7.4(a), "L," "R," and "B" denote left, right, and both, respectively.

In addition, to further evaluate the performance of our model, we compared our framework with 21 state-of-the-art methods which use the same data set. Specifically, we compared our model against 11 competitive state-of-the-art methods for motor imagery classification and against five cutting-edge methods for person identification and neurological diagnosis. Table 7.2 gives the comparison results.

We could observe that our proposed framework consistently outperformed a set of widely used baseline methods and strong competitors on three different data sets. The performance of our model thus shows a

Table 7.1: Comparison with baselines.

Scenarios	Data Sets	Metrics	Nondeep Learning Baselines					Deep Learning Baselines			
			SVM	RF	KNN	AB	LDA	LSTM	GRU	CNN	Ours
MIR	EEGMMIDB	Accuracy	0.5596	0.6996	0.5814	0.3043	0.5614	0.648	0.6786	0.91	**0.9632**
		Precision	0.5538	0.7311	0.6056	0.2897	0.5617	0.6952	0.8873	0.9104	0.9632
		Recall	0.5596	0.6996	0.5814	0.3043	0.5614	0.6446	0.6127	0.9104	0.9632
		F1-score	0.5396	0.6738	0.5813	0.2037	0.5526	0.6619	0.7128	0.9103	0.9632
PI	EEG-S	Accuracy	0.6604	0.9619	0.9278	0.35	0.6681	0.9571	0.9821	0.998	**0.9984**
		Precision	0.6551	0.9625	0.9336	0.3036	0.6779	0.9706	0.9858	0.998	0.9984
		Recall	0.6604	0.962	0.9279	0.35	0.6681	0.9705	0.9857	0.998	0.9984
		F1-score	0.6512	0.9621	0.9282	0.2877	0.668	0.9705	0.9857	0.998	0.9984
ND	TUH	Accuracy	0.7692	0.92	0.9192	0.5292	0.7675	0.6625	0.6625	0.9592	**0.9975**
		Precision	0.7695	0.9206	0.923	0.7525	0.7675	0.6538	0.6985	0.9593	0.9975
		Recall	0.7692	0.92	0.9192	0.5292	0.7675	0.6417	0.6583	0.9592	0.9975
		F1-score	0.7692	0.9199	0.9188	0.3742	0.7675	0.6449	0.6685	0.9592	0.9975

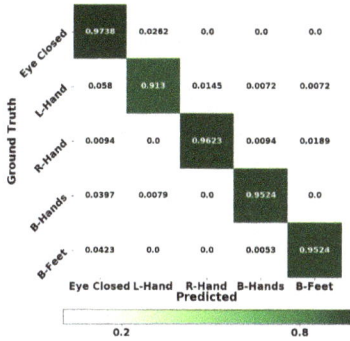

(a) CM of EEGMMIDB

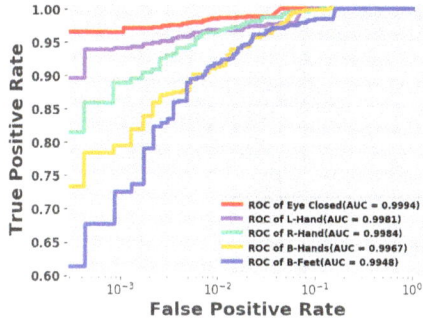

(b) ROC of EEGMMIDB

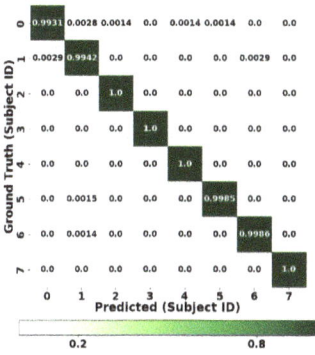

(c) CM of EEG-S

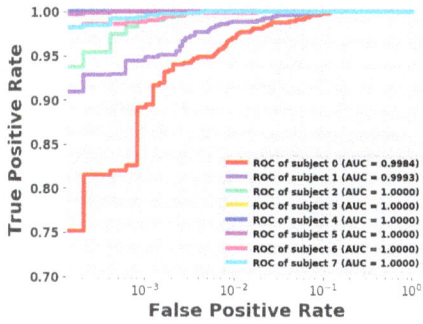

(d) ROC of EEG-S

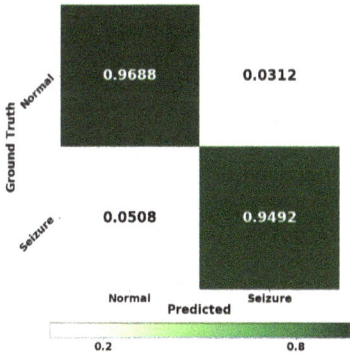

(e) CM of TUH

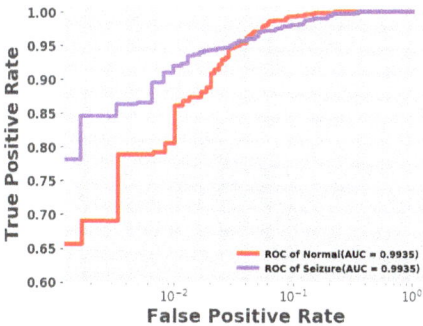

(f) ROC of TUH

Fig. 7.4: Confusion matrix and ROC curves with AUC scores of each data set. CM denotes confusion matrix.

Table 7.2: Comparison with the state-of-the-art approaches.

Scenarios	Data Sets	Metrics	State-of-the-Art					
MIR	EEGMMIDB	Method	or Rashid and Ahmad (2016)	Zhang et al. (2018h)	Ma et al. (2018)	Alomari et al. (2014a)	Sita and Nair (2013)	Alomari et al. (2014b)
		Accuracy	0.9193	0.9561	0.6820	0.8679	0.7584	0.8515
		Precision	0.9156	0.9566	0.6971	0.8788	0.7631	0.8469
		Recall	0.9231	0.9621	0.7325	0.8786	0.7702	0.8827
		F1-score	0.9193	0.9593	0.7144	0.8787	0.7666	0.8644
		Method	Shenoy et al. (2015)	Szczuko (2017)	Stefano Filho et al. (2017)	Pinheiro et al. (2016)	Kim et al. (2016)	Ours
		Accuracy	0.8308	0.9301	0.8724	0.8488	0.8115	**0.9632**
		Precision	0.8301	0.9314	0.8874	0.8513	0.8128	0.9632
		Recall	0.8425	0.9287	0.8874	0.8569	0.8087	0.9632
		F1-score	0.8363	0.9300	0.8874	0.8541	0.8107	0.9632
PI	EEG-S	Method	Ma et al. (2015)	Yang and Deravi (2014)	Rodrigues et al. (2016)	Fraschini et al. (2015)	Thomas and Vinod (2016a)	Ours
		Accuracy	0.88	0.99	0.8639	0.956	0.9807	**0.9984**
		Precision	0.8891	0.9637	0.8721	0.9458	0.9799	0.9984
		Recall	0.8891	0.9594	0.8876	0.9539	0.9887	0.9984
		F1-score	0.8891	0.9615	0.8798	0.9498	0.9843	0.9984
ND	TUH	Method	Ziyabari et al. (2017)	Harati et al. (2015)	Zhang et al. (2018j)	Goodwin and Harabagiu (2017)	Golmohammadi et al. (2017b)	Ours
		Accuracy	0.9382	0.9429	0.994	0.924	0.9479	**0.9975**
		Precision	0.9321	0.9503	0.9951	0.9177	0.9438	0.9975
		Recall	0.9455	0.9761	0.9951	0.9375	0.9522	0.9975
		F1-score	0.9388	0.9630	0.9951	0.9275	0.9480	0.9975

significant improvement over other baselines. These data sets were collected using a variety of EEG hardware, ranging from high-precision medical equipment to off-the-shelf EEG headsets with differing numbers of EEG channels. Regarding the seizure diagnosis in ND, by setting the normal state as impostor and the seizure state as genuine, our approach achieved a false acceptance rate (FAR) of 0.0033 and a false rejective rate (FRR) of 0.0017, outperforming the existing methods by a large margin (Acar *et al.*, 2007; Goodwin and Harabagiu, 2017; Golmohammadi *et al.*, 2017b; Harati *et al.*, 2015).

7.3.4 *Discussion*

This section proposes a generic and effective framework for raw EEG signal classification to support the development of several brain signal applications. The framework works directly on raw EEG data without requiring any preprocessing or feature engineering. Moreover, it can automatically select distinguishable feature dimensions for different EEG samples, thus achieving high usability across applications.

7.4 Transfer Learning Methods

We can build cross-scenario BCI systems by taking advantage of the power of transfer learning. Transfer learning methods acquire general knowledge from a source scenario and apply this learning to a new target scenario. In this way, a well-designed BCI system can handle several applications.

To the best of our best knowledge, there has been no research as yet which has combined deep learning and transfer learning for scenario adaption in the BCI area. However, in this section, we shed some light on this topic by presenting a cross-scenario transfer model based on traditional machine learning models (i.e., SVM) (Bruzzone and Marconcini, 2009).

Bruzzone *et al.* explore the problem that only the source scenario samples are labeled; the samples from the target scenario are unlabeled. The former are used in the training stage, while the latter are used in the testing stage. The insight behind transfer learning is that the sample distribution of the source scenario can be used to improve performance in the target scenario because the two scenarios are correlated. During the training stage, the samples from the source scenario are only utilized to initialize the discriminant function used for prediction. Then the samples are successively removed so that at last the final hyperplane is only defined on the basis of samples from the target scenario. In other words, the proposed model carefully and iteratively eliminates the source scenario instances one by one. The discriminant function is consequently adapted to the target scenario samples, which can recover distinctive information and properly address the target scenario classification problem.

Let E_s and E_t denote the data set of the source scenario (associated with labels Y_s) and the target scenario (without labels). The cross-scenario algorithm contains three stages: initialization, iterative scenario adaption, and convergence. In the first stage, we calculate a separation hyperplane only by E_s, as a standard SVM. Afterward, we use the above hyperplane to classify E_t and assign them labels Y_t. Obviously, the initial hyperplane performs erratically. To improve the separation hyperplane, we iteratively select a subset of E_t, named E'_t that has the highest distance to the hyperplane. The size of the E'_t is a user-defined hyperparameter. At the same time, a subset from E_s, named E'_s that has the lowest distance to the hyperplane, is also selected. Then we update the separation hyperplane by the union of E'_t and E'_s. Repeat the first two stages until the stop trigger (as introduced in the last stage) is met. In the last stage, the algorithm stops iterating while both the number of mislabeled and remaining unlabeled samples lying in the margin band at the current iteration is lower than or equal to a predefined threshold value.

This algorithm is based on an SVM classifier and can be easily extended to deep learning models. In such a cross-scenario model, labeled brain signals from the source scenario are employed to build an initial classifier. Then the unlabeled samples from the target scenario are utilized to adjust the decision boundary in order to adapt the classifier to the target scenario problem. To some extent, this idea is similar to the principle of the k-means cluster. Although the computation cost of this model is quite high since it is necessary to retrain the classifier at each iteration, it still suggests a promising direction for future research.

Chapter 8

Semi-Supervised Classification

In most real-world scenarios, accurate annotation of the collected brain signals is time-consuming. In this chapter, we therefore investigate semi-supervised classification in BCI systems where the distribution of the labeled data can be enhanced by the distribution of the unlabeled observations.

Semi-supervised learning, lying conceptually between supervised and unsupervised learning, is one of the fundamental challenges in the field of machine learning; it concerns the difficulties which arise when only a subset of the observations has corresponding class labels (Ghasedi Dizaji *et al.*, 2018). This issue is of immense practical interest in a broad range of application scenarios, such as abnormal activity detection (Yao *et al.*, 2016), neurological diagnosis (Peng *et al.*, 2016), and computer vision (Gong *et al.*, 2016). In such scenarios, observations are easily obtained in abundance, but it is expensive to obtain the corresponding class labels.

Based on how they learn information from unlabeled samples, the semi-supervised classification methods can be divided into three categories: generative methods, wrapper methods, and unsupervised representation learning (Van Engelen and Hoos, 2020). Next, we present the background on these and provide insights into the corresponding algorithms.

8.1 Generative Methods

8.1.1 *Overview*

The generative methods are able to generate a batch of new samples based on the distribution of unlabeled samples. In this way, the models can help to solve semi-supervised classification by augmenting the training data set. Among existing generative approaches, variational autoencoders (VAEs) (Kingma *et al.*, 2014; Sønderby *et al.*, 2016) have recently achieved state-of-the-art performance in semi-supervised learning.

VAE models provide a general framework for learning latent representations: a model is specified by a joint probability distribution both over the data and over latent random variables, and a representation can be found by considering the posterior on latent variables given specific data (Narayanaswamy *et al.*, 2017). The learned representations can be used not only for generation but also for classification. For instance, VAE provides a latent feature representation of the input observations; a separate classifier can be trained thereafter using these same representations. The high quality of latent representations enables accurate classification, even with a limited number of labels. A number of studies have applied VAE in semi-supervised classification in the computer vision area (Kingma *et al.*, 2014; Makhzani *et al.*, 2015; Narayanaswamy *et al.*, 2017).

Why we propose the VAE++. One major challenge faced by VAE-based semi-supervised methods is that the latent representations are stochastically sampled from the prior distribution instead of being rendered directly from the explicit observations. In particular, as shown in Figure 8.1(a), the learned latent representations z_s are *randomly sampled* from a multivariate Gaussian distribution (Equation (8.1)). Thus, to provide a specific sample, the corresponding latent representation is not exclusive (i.e., the representation is not repeatable in different runnings), which makes it inappropriate for classification. To solve this problem, in the latent space, we propose a new variable z_I (Figure 8.1(b)) which is directly

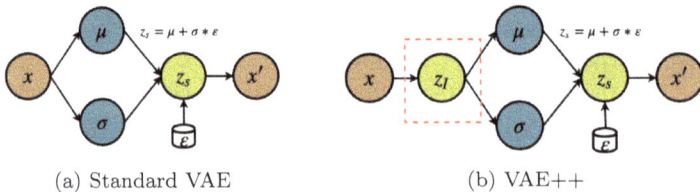

(a) Standard VAE   (b) VAE++

Fig. 8.1: Comparison of the standard VAE and the proposed VAE++. x and x' denote the input and the reconstructed data; μ and σ denote the learned expectation and standard deviation; z_s denotes the stochastically sampled latent representation, which is composed by μ, σ, and ε, where ε is randomly sampled from $\mathcal{N}(0, 1)$. In the standard VAE, z_s is regarded as the learned representation, while in VAE++, z_I denotes the proposed exclusive latent representation which can be used for classification.

learned from the input data. The exclusive latent code z_I is guaranteed to remain invariant for a specific input x in different runnings. The modified VAE is called VAE++.

In addition, the learned expectation μ contains only partial information on the input observations, which is not enough to represent the observations in the classification task, even though μ is exclusive.[1] The performance comparison among z_I, z_s and μ will be presented in Section 8.1.3.

Why VAE++ needs the semi-supervised GAN. In the proposed VAE++, it is necessary to reduce the information loss between the two latent representations z_I and z_s to guarantee the learned z_I is representative. The commonly used constraints between two distributions (e.g., Kullback–Leibler divergence) can only utilize the information of the observations they fail to exploit the information of labels. In this chapter, we adopt a novel approach that will allow us to take advantage of both unlabeled and labeled data by jointly training the VAE++ and a semi-supervised GAN.

Why a semi-supervised GAN needs the VAE++. GAN-based approaches (Odena, 2016; Salimans *et al.*, 2016) have shown promising results in semi-supervised learning. The semi-supervised GAN trains a generator and a discriminator with inputs belonging to one of K classes. In contrast to the regular GAN, the semi-supervised GAN requires the discriminator to make a $K + 1$ class prediction with an extra class added, corresponding to the generated fake samples. In this way, the observations' properties can be used to improve decision boundaries and allow for more accurate classification than would be achieved by than using the labeled data alone. However, the generated samples are taken from predefined distribution (e.g., Gaussian noise) (Cao *et al.*, 2018). Such predefined prior distributions are often independent from the input data distributions and obstruct the convergence and therefore cannot guarantee the distribution of the generated data. This drawback can be addressed by gearing with VAE++ which can provide a meaningful prior distribution that represent the distribution of the input data.

We introduce a recipe for semi-supervised learning, a robust adversarial variational embedding (AVAE) framework, which learns the exclusive latent representations by combining VAE and semi-supervised GAN. To utilize

---

[1]For the same reason, σ cannot be used as the exclusive code.

the generative ability of GAN and the distribution-approximating power of VAE, the proposed approach employs GAN to encourage VAE with the aim of learning the more robust and informative latent code. We present the framework in the context of VAE, adding a new exclusive code in the latent space which is directly rendered from the data space. The generator in VAE++ also works as a generator of GAN-both the exclusive code (marked as real) and the generated representation (marked as fake) are fed into the discriminator in order to force them to produce similar distributions (Mirza and Osindero, 2014).

8.1.2 *Adversarial Variational Embedding Algorithm*

Suppose the input data set has two subsets, one of which contains labeled samples, while the other contains unlabeled samples. In the former subset, the observations appear as pairs $(\boldsymbol{X}^L, \boldsymbol{Y}^L) = \{(\boldsymbol{x}_1^L, \boldsymbol{y}_1),$ $(\boldsymbol{x}_2^L, \boldsymbol{y}_2), \ldots, (\boldsymbol{x}_{N_L}^L, \boldsymbol{y}_{N_L})\}$ with the ith observation $\boldsymbol{x}_i^L \in \mathbb{R}^M$ and the corresponding one-hot label $\boldsymbol{y}_i \in \mathbb{R}^K$ where K denotes the number of classes, N_L denotes the number of labeled observations, and M denotes the number of the observation dimensions. In the latter subset, only the observations $\boldsymbol{X}^U = \{\boldsymbol{x}_1^U, \boldsymbol{x}_2^U, \ldots, \boldsymbol{x}_{N_U}^U\}$ are available, and N_U denotes the number of unlabeled observations $\boldsymbol{x}_i^U \in \mathbb{R}^M$. The total data size N equals to the sum of N_L and N_U. In terms of effective classification, we attempt to learn a latent representation with distinguishable information. Then the learned representations can be fed into a classifier for recognition. In this chapter, we mainly focus on the latent code learning.

In semi-supervised learning, because of the lack of labeled observations, it is necessary to learn latent variable distribution based on the observations without labels.[2] Thus, an encoder must be built to provide an embedding or feature representation which allows accurate classification even with limited observations.

8.1.2.1 *VAE++*

The VAE has been demonstrated to provide a latent feature representation for semi-supervised learning (Kingma *et al.*, 2014; Narayanaswamy *et al.*, 2017), as opposed to a linear embedding method or a regular autoencoder.

---

[2] For simplification, we omit the index and directly use variable $\boldsymbol{x}$ to denote observations.

The VAE maps the input observation x to a compressed code z_s, and then decodes it to reconstruct the observation. The latent representation is computed through the reparameterization trick (Kingma and Welling, 2013):

$$z_s = \mu_x + \sigma_x * \varepsilon \qquad (8.1)$$

with $\varepsilon \sim \mathcal{N}(0,1)$ to impose the posterior distribution of the latent code on $p(z_s|x) \sim \mathcal{N}(\mu_x, \sigma_x^2)$. μ_x and σ_x denote the expectation and standard deviation of the posterior distribution of z_s, which are learned from x. For the efficient generation and reconstruction, VAE imposes the code z_s on a prior Gaussian distribution:

$$\bar{p}(z_s) = \mathcal{N}(z_s|\mathbf{0}, \mathbf{I}) \qquad (8.2)$$

Through minimizing the reconstruction errors between x and x' and restricting the distribution of z_s to approximate the prior distribution $\bar{p}(z_s)$, VAE is expected to learn the representative latent code z_s, which can be used for both classification and generation.

Due to its strong feature representation ability, VAE has been employed for feature extraction and semi-supervised learning (Narayanaswamy *et al.*, 2017; Walker *et al.*, 2016). However, one limitation of the standard VAE is that the learned latent code $z_s = g(\mu_x, \sigma_x, \varepsilon)$, as shown in Equation (8.1), is not exclusive. In other words, for a specific observation x and a fixed embedding model $p(z_s|x)$, the corresponding latent code z_s is not exclusive as it contains a stochastic variable ε which is randomly sampled from the prior distribution $\bar{p}(z_s)$. For instance, in a pre-trained fixed VAE encoder, the specific input x will lead to a variety of z_s in different runnings. At a high level, the latent code z_s is determined by two factors: the prior distribution of observation $\bar{p}(x)$ which affects z_s through the learned μ_x and σ_x, and the stochastically sampled data ε. However, the stochastically sampled latent code is unstable and will corrupt the features for classification. Furthermore, the posterior distribution of z_s is forced to approximate the manually set prior distribution — commonly Normal Gaussian distribution — which inevitably leads to information loss.

In order to completely sidestep the abovementioned issue, we propose a novel VAE++ model to learn an exclusive latent code z_I. The VAE++ contains three key components: the encoder, the generator, and the decoder (see Figure 8.2). The encoder transforms the observation into a latent code

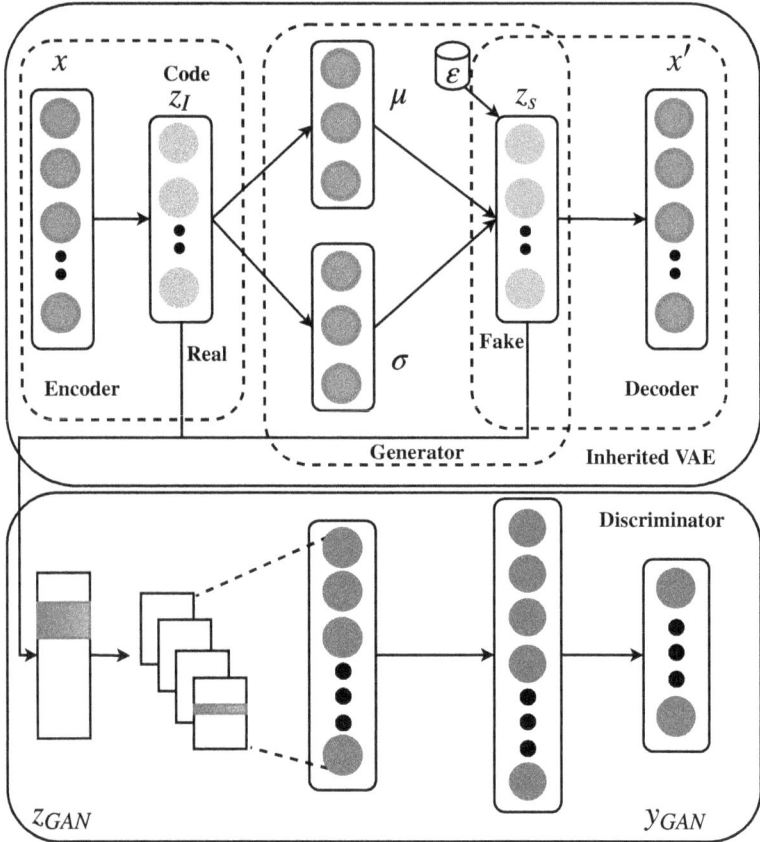

Fig. 8.2: An AVAE is composed of a VAE++ and a semi-supervised GAN. The generated z_s (labeled as fake) and the exclusive code z_I (labeled as real) are fed into the discriminator. The discriminator can exploit both the labeled and the unlabeled observations. The generator in the VAE++ also works as a generator of the GAN.

$z_I \in \mathbb{R}^D$ which is directly determined by the input x. D denotes the dimension of z_I. We learn:

$$p_{\theta_{en}}(z_I|x) = f(z_I; x, \theta_{en}) \tag{8.3}$$

where f denotes a nonlinear transformation while θ_{en} denotes the encoder parameters. The nonlinear transformation f is generally chosen as a deep neural network for its excellent ability to perform nonlinear approximation. Then, in the generator, we measure the expectation $\mu(z_I)$ and the standard derivation $\sigma(z_I)$ from the latent code z_I and update Equation (8.1).

The generated variable z_s can be calculated by:

$$z_s = \mu(z_I) + \sigma(z_I) * \varepsilon \tag{8.4}$$

At last, the decoder is employed to reconstruct the sample:

$$p_{\theta_{de}}(x'|z_s) = f'(x'; z_s, \theta_{de}) \tag{8.5}$$

where f' denotes another nonlinear rendering, called the decoder, with parameters θ_{de} and x' denotes the reconstructed observation.

The loss function of VAE++ can be calculated by:

$$\mathcal{L}_{VAE} = -\mathbb{E}_{z_s \sim p_{\theta_{en}}(z_s|x)}[\log p_{\theta_{de}}(x'|z_s)]$$
$$+ KL(p_{\theta_{en}}(z_s|x)||\bar{p}(z_s)) \tag{8.6}$$

The first component is the reconstruction loss, which equals to the expected negative log-likelihood of the observation. This term encourages the decoder to reconstruct the observation x based on the sampling code z_s which is under Gaussian distribution. The lower reconstruction error indicates that the encoder has learned a better latent representation. The second component is the Kullback–Leibler divergence, which measures the distance between the prior distribution of the latent code $\bar{p}(z_s)$ and the posterior distribution $p(z_s|x)$. This divergence reflects the information loss when we use $p(z_s|x)$ to represent $\bar{p}(z_s)$.

In the latent space of the novel VAE++, there are two compressed informative codes, z_I and z_s. The former represents directly encoded x, whereas the latter is stochastically sampled from the posterior distribution, which makes the former more suitable for classification. Therefore, we choose z_I as the compressed latent code in VAE++ instead of z_s, as used in the standard VAE.

From Equation (8.4), we can observe that the expectation and standard deviation of z_s and z_I are invariant. In particular, for a specific sample x_i, the corresponding z_{si} and z_{Ii} have the same statistical characteristics. Thus, we have

$$z_s \leftarrow \mu(z_I), \sigma(z_I), \varepsilon \tag{8.7}$$

which indicates that the generated z_s is affected by both the distribution of z_I and the prior distribution $\bar{p}(z_s)$ (or ε). In summary, the z_s inherits the statistical characteristics of z_I.

8.1.2.2 *Adversarial Variational Embedding*

One significant sufficient condition of a well-trained VAE++ is less information loss in the transformation from z_I to z_s to guarantee that the learned z_I is representative. As previously mentioned, part of the information in z_s is inherited from z_I and the other part is randomly sampled from the prior distribution $\bar{p}(z_s)$. Since the conditional distribution $p_{\theta_{en}}(z_I|x)$ has a better description of the input observation x, we increase the proportion of the inherited part and decrease the proportion of the stochastically sampled part.

As shown in Figure 8.2, in the proposed AVAE, the generator G generates z_s based on the joint probability $p(\mu, \sigma, \bar{p}(z_s))$ rather than the noise in the standard GAN. The z_s is regarded as "fake" while z_I is regarded as "real." Specifically, for the labeled observations x^L, VAE++ encodes the input to the latent code $z_I^L \in \mathbb{R}^D$ and generates $z_s^L \in \mathbb{R}^D$; similarly, for unlabeled observations x^U, we have $z_I^U \in \mathbb{R}^D$, which generates $z_s^U \in \mathbb{R}^D$. To exploit the label information, we extend the $y \in \mathbb{R}^K$ which has K possible classes to $y_{GAN} \in \mathbb{R}^{K+1}$ which has $K + 1$ possible classes by regarding the generated fake samples z_s as the $(K+1)$th class (Odena, 2016; Salimans *et al.*, 2016). In the VAE++, the unspecified z_s denotes both z_s^L and z_s^U whenever we don't care whether the observation is labeled or not. This rule also applies to z_I. Similarly, we use z_{GAN} to denote the input of the discriminator D, which contains both z_I and z_s. The discriminator can be described by

$$q_\varphi(y_{GAN}|z_{GAN}) = h(y_{GAN}; z_{GAN}, \varphi) \tag{8.8}$$

where φ denotes the parameters of D while h denotes the nonlinear transformation which is implemented by a CNN (Krizhevsky *et al.*, 2012). Therefore, we can use $q_\varphi(y_{GAN} = K + 1|z_{GAN})$ to supply the probability where z_{GAN} is fake (from z_s) and use $q_\varphi(y_{GAN}|z_{GAN}, y_{GAN} < K + 1)$ to supply the probability where z_{GAN} is real (from z_I) and is correctly classified.

For the labeled input it is the same as supervised learning: the discriminator not only tells us whether the input z_{GAN} is real or generated, but also sorts it into the correct class. Therefore, we have the supervised loss function

$$\mathcal{L}_{label} = -\mathbb{E}_{z_{GAN}, y_{GAN} \sim p_j}[log q_\varphi(y_{GAN}|z_{GAN}, y_{GAN} < K + 1)] \tag{8.9}$$

where p_j denotes the joint probability.

For the unlabeled input, we only require the discriminator to perform a binary classification: the input is real or fake. The former probability can be calculated by $1 - q_\varphi(\boldsymbol{y}_{GAN} = K + 1|\boldsymbol{z}_{GAN})$, whereas the latter can be calculated by $q_\varphi(\boldsymbol{y}_{GAN} = K + 1|\boldsymbol{z}_{GAN})$. Thus, the unsupervised loss function:

$$\mathcal{L}_{unlabel} = -\mathbb{E}_{\boldsymbol{z}_{GAN} \sim p_{\theta_{en}}(\boldsymbol{z}_I|\boldsymbol{x})}[log(1 - q_\varphi(\boldsymbol{y}_{GAN} = K + 1|\boldsymbol{z}_{GAN}))]$$
$$- \mathbb{E}_{\boldsymbol{z}_{GAN} \sim p_{\theta_{en}}(\boldsymbol{z}_s|\boldsymbol{x})}[log(q_\varphi(\boldsymbol{y}_{GAN} = K + 1|\boldsymbol{z}_{GAN}))]$$

In summary, the final loss function of the discriminator

$$\mathcal{L}_{GAN} = w_1 * flag * \mathcal{L}_{label} + w_2 * (1 - flag) * \mathcal{L}_{unlabel} \tag{8.10}$$

where w_1, w_2 are weights and $flag$ is a switch function

$$flag = \begin{cases} 1 & labeled \\ 0 & unlabeled \end{cases}$$

If the specific observation is labeled, we calculate the labeled loss function. Otherwise, we calculate the unlabeled loss function. From empirical experiments, we observe that the $\mathcal{L}_{unlabel}$ is much easier to converge than $\mathcal{L}_{label}$ and that the real/fake classification accuracy is much higher than the classification accuracy K classes. To encourage the optimizer to focus on the former part, which is more difficult to converge, we set $w_1 = 0.9$ and $w_2 = 0.1$.

The discriminator receives $\boldsymbol{z}_{GAN}$ as input and extracts the dependencies through CNN filters. Two fully connected layers follow the convolutional layer for dimension reduction. Finally, a softmax layer is employed to work on the low-dimensional features to estimate the log normalization of the categorical probability distribution, which is output as $\boldsymbol{y}_{GAN}$.

The overall aim of the proposed AVAE (as described in Algorithm 3) is to train a robust and effective semi-supervised embedding method. The VAE loss $\mathcal{L}_{VAE}$ and the GAN loss $\mathcal{L}_{GAN}$ are trained simultaneously by the Adam optimizer. After convergence, the compressed representative code $\boldsymbol{z}_I$ is fed into a nonparametric nearest neighbors classifier for recognition.

8.1.3 *Evaluation*

Next, we demonstrate the effectiveness and validation of the proposed method on neurological disorder diagnosis.

EEG signals collected from an individual in the unhealthy state differ significantly from those collected from an individual in the normal state (Adeli *et al.*, 2007). The Epileptic is a common brain disorder that affects about 1% of the population and its octal state can be detected by EEG

Algorithm 3 Adversarial Variational Embedding
___
Input: Labeled observations $(\boldsymbol{X}^L, \boldsymbol{Y}^L)$ and unlabeled observations $\boldsymbol{X}^U$
Output: Representation $\boldsymbol{z}_I$
  1: Initialize network parameters $\boldsymbol{\theta}_{en}, \boldsymbol{\theta}_{de}, \boldsymbol{q}_{\varphi}$
  2: **for** $\boldsymbol{x} \in \{\boldsymbol{X}^L, \boldsymbol{X}^U\}$ **do**
  3:     $\boldsymbol{z}_I \leftarrow \boldsymbol{x}$
  4:     $\boldsymbol{\mu}, \boldsymbol{\sigma} \leftarrow \boldsymbol{z}_I$
  5:     Sampling $\boldsymbol{\varepsilon}$ from $\mathcal{N}(\boldsymbol{0}, \boldsymbol{I})$
  6:     $\boldsymbol{z}_s = \mu(\boldsymbol{z}_I) + \sigma(\boldsymbol{z}_I) * \boldsymbol{\varepsilon}$
  7:     $\boldsymbol{x}' \leftarrow \boldsymbol{z}_s$
  8:     $\mathcal{L}_{VAE} \leftarrow \boldsymbol{x}, \boldsymbol{x}', p(\boldsymbol{z}_s|\boldsymbol{x})$
  9:     **for** $z_I, z_s, \boldsymbol{y} \in \boldsymbol{Y}^L$ **do**
10:         $\boldsymbol{y}_{GAN} \leftarrow z_I, z_s$
11:         $\mathcal{L}_{GAN} \leftarrow \boldsymbol{y}_{GAN}, \boldsymbol{y}$
12:     **end for**
13:     Minimize $\mathcal{L}_{VAE}$ and $\mathcal{L}_{GAN}$
14: **end for**
15: **return** $\boldsymbol{z}_I$

---

analysis. In this application, we evaluate our framework using raw EEG data to diagnose epileptic seizure.

The approach is evaluated over the benchmark data set TUH (Obeid and Picone, 2016) as described in Section 7.3.3.2.

8.1.3.1   *Results and Discussion*

From Table 8.1, we can see that our approach outperforms all of the competitive baselines on the TUH data set. For instance, at a 60% supervision level, the proposed approach achieves the highest accuracy of 95.21%, which represents an improvement of around 4% over the other methods. The corresponding confusion matrix (Figure 8.3) infers that the normal state has higher accuracy than the seizure state. One possible reason is that the start and end stage of the seizure share similar characteristics with the normal state, which may lead to misclassification.

8.1.3.2   *Further Analysis*

Supervision Rate. We have conducted extensive experiments to investigate the impact of supervision rate $\boldsymbol{\lambda}$. The supervision rate ranges from

Table 8.1: Overall comparison of semi-supervised classification accuracy (%) on neurological diagnosis.

Data Set	Rate (%)	Algorithm-Related State-of-the-Art			
		M2	AAE	LVAE	ADGM
Neurological Diagnosis (TUH)	20	71.28 ± 0.16	80.13 ± 0.95	82.31 ± 0.19	86.32 ± 0.12
	40	75.32 ± 0.16	82.95 ± 0.26	84.38 ± 0.16	86.99 ± 0.05
	60	76.32 ± 0.29	86.21 ± 0.52	87.51 ± 0.26	87.65 ± 0.16
	80	79.65 ± 0.37	88.53 ± 0.28	89.56 ± 0.25	88.05 ± 0.12
	100	82.59 ± 0.31	89.58 ± 0.25	90.25 ± 0.21	88.65 ± 0.26

Data Set	Rate (%)	Application-Related State-of-the-Art			
		Ziyabari et al. (2017)	Harati et al. (2015)	Schirrmeister et al. (2017)	Goodwin and Harabagiu (2017)
Neurological Diagnosis (TUH)	20	87.66 ± 0.23	86.38 ± 0.36	82.19 ± 0.24	86.33 ± 0.21
	40	89.25 ± 0.19	91.58 ± 0.35	84.21 ± 0.08	89.25 ± 0.34
	60	91.28 ± 0.37	92.58 ± 0.26	85.36 ± 0.32	90.38 ± 0.24
	80	92.59 ± 0.26	93.25 ± 0.31	85.16 ± 0.24	91.59 ± 0.16
	100	93.32 ± 0.18	94.29 ± 0.25	86.42 ± 0.26	92.4 ± 0.25

Data Set	Rate (%)	Ablation Study			Ours
		VAE (μ)	VAE	VAE++	AVAE
Neurological Diagnosis (TUH)	20	80.58 ± 0.69	86.37 ± 0.24	0.86 ± 0.53	93.69 ± 0.16
	40	81.35 ± 0.24	89.69 ± 0.27	91.28 ± 0.25	94.32 ± 0.28
	60	82.59 ± 0.63	90.58 ± 0.27	92.87 ± 0.31	95.21 ± 0.21
	80	83.21 ± 0.21	91.69 ± 0.35	93.96 ± 0.28	97.86 ± 0.26
	100	84.21 ± 0.65	92.38 ± 0.41	94.65 ± 0.24	98.13 ± 0.32

Note: If the compared method cannot deal with unsupervised samples, it will be trained only on supervised samples.

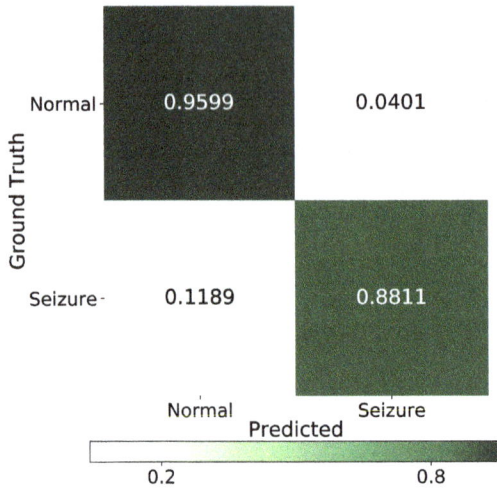

Fig. 8.3: Confusion matrix of TUH.

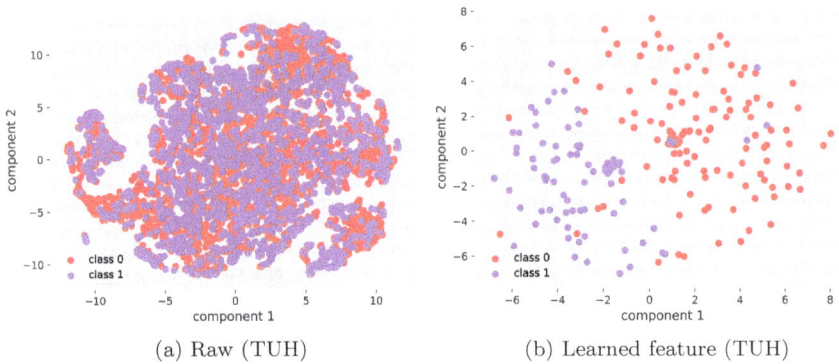

(a) Raw (TUH)   (b) Learned feature (TUH)

Fig. 8.4: Visualization for comparison between raw data and the semi-supervised learned features.

20% to 100% with a 20% interval and each setting runs at least three times with the average accuracy recorded. From Table 8.1, it can be seen that the proposed model attains a competitive performance at each supervision level.

Visualization. Figure 8.4 visualizes the raw data and the learned features of different data sets. The visualization comparison demonstrates the capacity of our approach for feature learning.

8.1.4 *Discussion*

First, the AVAE requires adequate labeled training samples. Through from supervision rate analyze results, the proposed approach demands a sufficient supervised proportion. Attaining a lower supervision rate should be one major goal for the future.

Moreover, the AVAE partially depends on the hyperparameter settings. The hyperparameter tuning (not presented here in detail due to space limitation), especially the key parameters like learning rate, is necessary for different data types. In our experiments, we chose one setting for activation recognition and the other setting for EEG signal recognition. For the parameter tuning, the orthogonal array experiment method is suggested (Zhang *et al.*, 2017b). A more generalized framework which is not sensitive to data types would be one future scope.

Furthermore, the classification accuracy of the discriminator in GAN is not excellent, even though the overall framework achieves a competitive performance. In the discriminator, the unlabeled classification (fake or real) achieves an accuracy of almost 100%. However, for the labeled observations, the classification accuracy ($K+1$ classes) is much lower. We believe this is a meaningful and crucial future direction for research in that an end-to-end semi-supervised classification framework could be built if the discriminator were to achieve satisfactory performance.

8.2 Wrapper Methods

The wrapper methods refer to the semi-supervised learning procedures of converting unlabeled samples to pseudo-labeled samples. Specifically, the assumption underlying wrapper models is that a label (called a pseudo-label) is assigned to the unlabeled sample through a classifier trained on the original labeled samples, and the decision boundary is updated by considering the pseudo-labeled samples.

The wrapper methods generally contain three stages: classifier training, pseudo-labeling, and data set update. First, a supervised classifier, called the base classifier is trained on the labeled data. Then the trained classifier is used to infer the label for each unlabeled sample. To this end, every unlabeled sample is given a pseudo-label. Third, a subset of the pseudo-labeled samples that have high confidence is selected and added to the labeled data set, which will be used to update the supervised classifier. The three stages are performed iteratively until a predefined convergence criterion is satisfied. In the end, we have a well-trained classifier which is then used to make the final prediction for testing samples.

Next, we present the subcategories of wrapper methods, including self-training, co-training, and boosting methods.

8.2.1 *Self-Training*

Self-training models, as the most basic approaches, refer to the wrapper methods that only employ a single base classifier (Li *et al.*, 2008). There is only one classifier and one data set during training. The model iteratively updates the pseudo-labels to enhance the decision boundary of the classifier itself, which is why it is called "self-training."

Next, we briefly provide some ideas on how to combine a self-training strategy with the popular deep learning classifiers (such as the representation learning methods in Chapter 6). For example, we build a CNN classifier to extract spatial features from EEG signals for binary emotion recognition (positive versus negative). Suppose we have 1000 labeled EEG segments E_l, 9000 unlabeled EEG segments E_u, and 1000 testing samples. Assuming the training set is balanced, let's look at how to perform self-training for semi-supervised emotional classification.

To begin with, we build a CNN classifier f and train it on E_l. In the classifier training, we select 80% of our samples from E_l as the training set and the residual as the validation set. The validation set will be used to calculate the classification error in order to perform backpropagation. Within the training of f, CNN converges based on the early stopping criterion.

We use the trained f to predict the pseudo-labels of E_u. Define the highest prediction probability at the output layer of CNN as the prediction confidence, then calculate the prediction confidence, then calculate of E_u and sort them in ascending order and select the 500 samples from E_u with the highest prediction confidence. To this end, the original E_u is split into one subset of 500 samples (called E_{ul}^1) and another subset with the remaining 8,500 pseudo-samples (called E_{uu}^1).

Third, we update the labeled data set as E_l^1 by combining E_l and E_{ul}^1. Please note that at this point, there are 1500 labeled sample E_l^1 for which we have high confidence that the labels are correct. We have 8500 pseudo-samples E_{uu}^1 for which we don't trust the pseudo-labels.

Let's move to the second iteration. Similar to the first round, we train a classifier f^1 on $\boldsymbol{E}_l^1$. Here, we assume f^1 is more powerful and generalized than f because it considers the information extracted from $\boldsymbol{E}_{ul}^1$. Second, we infer the samples of $\boldsymbol{E}_{uu}^1$ by f^1, select the best 500 samples $\boldsymbol{E}_{ul}^2$ and leave the residual 8000 samples ($\boldsymbol{E}_{uu}^2$). Next, we introduce the principle of "survival of the fittest" by removing some unfit samples from $\boldsymbol{E}_l^2$ — specifically, we use f^1 to predict the labels of $\boldsymbol{E}_l^1$ and select 100 samples with the lowest prediction confidence and remove them. The remainder of the data set is called $\bar{\boldsymbol{E}}_l^1$, which combined with $\boldsymbol{E}_{ul}^2$ to form the new labeled data set $\boldsymbol{E}_l^2$. Note that $\boldsymbol{E}_l^2$ has $1900 = 1000 + 500 - 100 + 500$ samples.

The above operations are repeated until the size of $\boldsymbol{E}_l^N$ is greater than 5000, where N denotes the number of iterations. Finally, we use the re-trained classifier f^N to predict the test samples.

8.2.2 Co-Training

The co-training methods are expanded from self-training in that the former contain more than one base classifier. The base classifier screens pseudo-samples with high prediction confidence and inserts them into the training set of the other classifier. The strategy of co-training works where the base classifiers are not strongly correlated. In other words, the different classifiers learn different aspects of the data set so that they can exchange information with each other and to improve prediction performance.

For a deeper understanding, let's continue to take the example in Section 8.2.1 and suppose there are two base classifiers: an RNN classifier named f^r and a CNN classifier named f^c.

All the settings are the same. In the first iteration, f_r and f_c are trained on $\boldsymbol{E}_l$. Then f^r predicts pseudo-labels for $\boldsymbol{E}_u$ and gets the $\boldsymbol{E}_{ul}^{r1}$ with the highest prediction confidence and $\boldsymbol{E}_{uu}^{r1}$ with low confidence. Similarly, for f^c, we have $\boldsymbol{E}_{ul}^{c1}$ and $\boldsymbol{E}_{uu}^{c1}$. Unlike in self-training, here, $\boldsymbol{E}_{ul}^{r1}$ is combined with $\boldsymbol{E}_l$ to form $\boldsymbol{E}_l^{c1}$ to improve the CNN classifier in the next iteration, whereas $\boldsymbol{E}_{ul}^{c1}$ is combined with $\boldsymbol{E}_l$ to form $\boldsymbol{E}_l^{r1}$ for the RNN classifier.

Due to the sample exchange, each base classifier can absorb information from its counterpart to finally boost its classification performance. This model can be easily expanded for multiple base classifiers such that each classifier can gather the high-confidence pseudo-samples from the remaining classifiers. Since multiple base classifiers are involved, the co-training strategy is generally used together with boosting (more details in Section 8.2.3).

8.2.3 *Boosting*

A wrapper method's performance can be boosted by combining a number of base classifiers. For the self-training strategy, we can independently train several classifiers on the input data. While predicting the test samples, all the prediction results from the independent base classifiers are considered through aggregation methods (e.g., voting or weighted average). Similarly, for co-training, the multiple base classifiers are trained dependently through the exchange of pseudo-samples. The prediction results of all the classifiers are combined to make the final decision.

In summary, wrapper methods use a very straightforward method of exploiting unlabeled data. On the one hand, one big advantage of wrapper models is that they can be easily incorporated into any existing supervised algorithms (deep learning or traditional classifiers). This is what makes the use of wrapper methods so flexible. For example, we can employ a single base classifier or several base classifiers; adopt a deep neural network–based classifier, a traditional classifier, or a mix of them; select different convergence demands; design different ways to measure the classification confidence; make different data set update rules, and so on. However, on the other hand, it is possible that a wrapper method may require a large number of labeled samples to initialize the base classifier. If there are too few labels for the initial training, the base classifier would be full of bias, infer wrong pseudo-labels, and ultimately fail at the semi-supervised prediction task. As we know, the wrapper methods have not attracted sufficient attention in deep learning-based BCI, and there are still a number of untapped opportunities for research in this direction.

8.3 Unsupervised Representations Learning

The union of unsupervised representative techniques with supervised classifiers does not strictly result in semi-supervised learning because they cannot exploit the information carried by unlabeled samples. However, it is listed as a semi-supervised approach in much of the literature (Jia *et al.*, 2014; Van Engelen and Hoos, 2020; Wulsin *et al.*, 2010). Following the peer-reviewed work, we consider a deep learning framework that contains both unsupervised and supervised components as semi-supervised learning.

In part, unsupervised feature extraction techniques are widely used in brain signal processing. In this section, we very briefly introduce this kind of semi-supervised learning in BCI scenarios. Algorithms with unsupervised preprocessing, such as PCA and wavelet transformation, are not considered.

As introduced in Section 3.2, there are a number of unsupervised representative deep learning algorithms (e.g., AE, RBM, and DBN) capable of purifying the raw brain signals, which have achieved great success in recent years. The representative deep neural networks are adopted to learn the latent representations from the high-dimensional raw data space to support the downstream tasks (generally a predictor).

Wulsin *et al.* (2010) shows that DBN can effectively learn representations for anomaly detection in EEG signals. Jia *et al.* (2014) proposes a semi-supervised deep learning framework which first employs a two-level channel selection component that learns representations on both labeled and unlabeled data, then uses an RBM-based model for classification, in which the training is regularized by both supervised and unsupervised information.

What we want to point out here is that the unsupervised representation learning is powerful for dimensionality reduction and latent feature discovery; however, it may also conflict with other deep learning models (such as an RNN classifier). For example, the unsupervised learned representation in the embedded space may lose the original temporal dependencies and thus result in a low-performance BCI. Consequently, researchers should carefully adopt or arrange the position of unsupervised representation networks when designing a deep learning framework for BCI systems.

PART 4

Typical Deep Learning
for EEG-Based BCI Applications

Chapter 9

Authentication

In this chapter, we present deep learning-based BCI applications for person identification and authentication, respectively. Person identification refers to the BCI system recognizing the identity of the user, whereas authentication refers to a binary classification in which the user is either authenticated or rejected.

9.1 EEG-Based Person Identification

9.1.1 *Challenges*

One of the most promising BCI applications is EEG-based user identification, which exploits the unique characteristics of EEG signals (such as resistance to fraud).

Over the past decade, biometric information has been widely adopted in identification and has gained in acceptance due to its reliability and adaptability. Existing biometric identification systems are based mainly on individuals' unique intrinsic physiological features (e.g., facial structure (Givens *et al.*, 2013), iris (Latman and Herb, 2013), retina (Sadikoglu and Uzelaltinbulat, 2016), voice (Ormerod, 2017), and fingerprint (Unar *et al.*, 2014)). However, even these state-of-the-art identification systems have been shown to be vulnerable. For example, anti-surveillance prosthetic masks can thwart face recognition, contact lenses can trick iris scanners, vocoder can compromise voice identification, and fingerprint films can deceive fingerprint sensors. In this environment, the EEG-based biometric identification systems are emerging as a promising alternative due to their high resilience to attack (Kerem and Geva, 2017; Sohankar *et al.*, 2015). EEG-based identification systems measure an individual's response to a number of stimuli in the form of EEG signals, which record the electromagnetic and invisible electrical neural oscillations. An individual's EEG signals are virtually impossible for an impostor to mimic, making this approach highly resistant to the type of spoofing attacks to which other identification techniques are vulnerable.[1]

EEG signals, compared with other biometrics, possess the following significant inherent advantages (Sohankar *et al.*, 2015):

- *Resilience to attack.* EEG data is invisible and untouchable, and thus impossible to clone or duplicate. Therefore, an EEG-based identification system is robust against faked identities.
- *Universality.* EEG signals are typically permanently associated with one specific subject; hence, security can be enforced anywhere and anytime.
- *Uniqueness.* Each individual's EEG signals are unique (Gui *et al.*, 2014). This can potentially allow a high degree of accuracy.
- *Accessibility.* We have seen increased effort in recent years devoted to the development of low-cost and easy-to-wear EEG headsets. For example, the behind-the-ear EEG collection equipment (Kidmose *et al.*, 2013) is easily attached to the ear (similar to wireless earphones).

Table 9.1 provides a comparison of EEG with other biometric information over several key characteristics.

However, research on EEG-based identification is still in its infancy, and several key challenges persist. One of the most significant issues is its low identification accuracy as a result of the inherent low precision of EEG signals; EEG data has a very low signal-to-noise ratio, making accurate identification challenging. The state-of-the-art approaches can achieve an accuracy in the range of 80% to 95% (Bashar *et al.*, 2016; Kerem and Geva, 2017; Thomas and Vinod, 2016b), which is not sufficient for practical deployment, particularly in high-security environments.

Another challenge is the poor stability — the identification system may work well in one instance but fail in another due to how sensitive EEG

---

[1]For example, people can easily trick a fingerprint-based identification system by using a fake fingerprint film (http://www.instructables.com/id/How-To-Fool-a-Fingerprint-Security-System-As-Easy-/. Accessed Sep. 15, 2016) or a face-recognition-based identification system by simply wearing a $200 anti-surveillance mask (http://www.urmesurveillance.com/urme-prosthetic/. Accessed Oct. 23, 2016).

Table 9.1: Comparison of various biometrics. EEG display considerable resilience to attack, which is one of the most significant characteristics of identification systems. ↑ denotes the higher the better, while ↓ denotes the lower the better.

Biometrics	Attack-Resilient ↑	Universality ↑	Uniqueness ↑	Stability ↑	Accessibility ↑	Performance ↑	Cost ↓
Finger/Palmprint	Low	High	High	High	Medium	High	Medium
Iris	Medium	High	High	High	Medium	High	High
Retina	High	Medium	High	Medium	Low	High	High
Signature	Low	High	Low	Low	High	Low	Medium
Voice	Low	Medium	Low	Low	Medium	Low	Low
Face	Medium	High	Medium	Medium	Medium	Medium	High
Gait	High	Medium	High	Medium	Medium	High	Low
EEG	**High**	High	High	Low	Medium	High	Low

signals are to interference. This may be caused by the user's physiological and psychological states such as fatigue and anger (Wang *et al.*, 2015a; Trejo *et al.*, 2015). Intuitively, the state shift brought by the fluctuation of user states can be divided into two categories: dramatic shifts (e.g., hysterical, drunk, or under threat) and slight shifts (e.g., headache or excitement). On the one hand, the EEG signal divergence introduced by the former can help to enhance the robustness of the identification system. For example, this phenomena could enhance the security of the system if it causes the system to fail to recognize a subject who is under threat (e.g., Hijacked abducted by a kidnapper). Knyazev (2012) infers that EEG signals are affected by inherent factors such as panic, sustained pain, and sexual arousal. On the other hand, however, the latter may reduce the signal quality but more commonly occur in the real world. For instance, the identification system might reliably identify the user when s/he is happy but fail to be able do so when s/he is upset. Thus, the slight shift should be overcome for its negative impact.

To address the aforementioned problems, we propose MindID, a delta pattern EEG-based person identification algorithm based on an attention-based RNN. First, to eliminate the interference of the slight shifts brought about by environmental noise and the varying physical and mental states of the subject, we attempt to learn robust and reliable representation by decomposing the EEG patterns. For this, we decompose the full spectrum of EEG data into specific patterns (delta, theta, alpha, beta, and gamma). Decomposed EEG patterns (e.g., theta, alpha, beta, and gamma), have been employed for EEG signal classification (e.g., movement task classification (Müller-Gerking *et al.*, 1999)) in some works. However, few existing studies have focused on the delta pattern. Through our analysis in Section 9.1.2, we have determined that the delta pattern is the most discriminative and efficient pattern. Moreover, we introduce the attention-based RNNs (Recurrent Neural Networks) (Bahdanau *et al.*, 2016) which can automatically detect the most distinctive information from the input EEG data. Most importantly, the attention mechanism automatically reallocates the weights to extract most discriminative features resilient to the change in environmental factors.[2] Therefore, the proposed approach is

---

[2]Simply put, the attention mechanism refers to the selection of the most pertinent piece of information rather than using all of the available information. Attention mechanisms in neural networks are based on the visual attention mechanism found in humans and have been applied in the computer version and natural language processing areas.

robust under different collection environments as well as resilient to changes to the EEG collection hardware, sampling rate, and channel numbers. Moreover, the efficiency of the attention-based RNN framework has already been demonstrated in some research fields — e.g., computer vision (Luong *et al.*, 2015).

9.1.2 *EEG Pattern Analysis*

In this section, we first introduce some background on EEG patterns, followed by a topographical analysis of real-world EEG data in order to discover which specific constituent patterns capture the most distinctive features and thus allow us to determine a subject's identity. Next, we analyze why the delta pattern works best, both qualitatively and quantitatively.

The EEG signals collected from any typical EEG hardware can be divided into several nonoverlapping frequency bands — delta, theta, alpha, beta, and gamma — based on the strong intraband correlation of each type with a distinct behavioral state (Section 2.2.1). Each decomposed EEG pattern contains signals associated with particular brain information. The EEG frequency patterns and their corresponding characteristics are listed in Table 2.2.

In order to investigate which EEG pattern is most intrinsic and rich in distinctive information for user identification, we have studied the EEG topography of the different frequency patterns, the data set for which was locally collected by our group. Figure 9.1 shows the EEG topography of various subjects on full bands — delta, theta, alpha, beta, and gamma patterns, respectively — illustrating that the delta pattern has the lowest intersubject similarity of all the patterns and thus is likely to offer the most distinctive features for person identification. Below, we present two arguments to explain why delta patterns are most suited for user identification. Prior studies have shown that the functional significance of delta oscillations is not yet fully understood (Knyazev, 2012); our arguments below are based on the current knowledge of these patterns.

On the one hand, qualitatively the delta pattern is universal and stable. It is a widely accepted view that the delta pattern occurs only in deep sleep states, which is a significant reason for why most researchers neglect the delta frequency in user identification. However, recent research in neurophysiology indicates that the delta rhythm is often evident during "quiet" wakeful states in rodents and nonhuman primates (Sachdev *et al.*, 2015). This suggests that the delta patterns may dominate the background

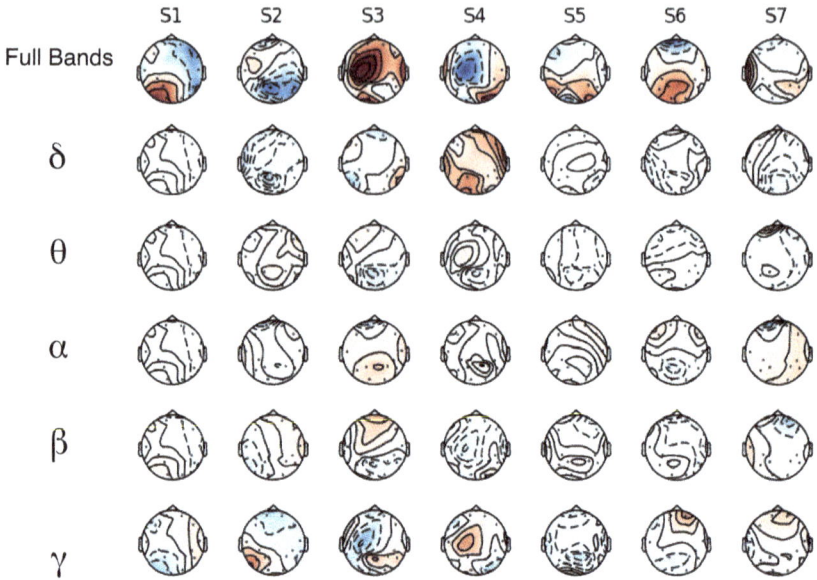

Fig. 9.1: EEG topography of various subjects under different frequency patterns. The intersubject EEG cosine similarity is calculated under each pattern and the results are reported as 0.1313 (full patterns), 0.0722 (delta), 0.1672 (theta), 0.2819 (alpha), 0.0888 (beta), and 0.082 (gamma). This illustrates that the delta pattern has the lowest intersubject similarity compared to other patterns and that it is thus likely to offer the most distinct features for person identification.

activity of some neocortical circuits in individuals in the waking state. In addition, the delta pattern has been observed to be related to cognitive processing (Harmony, 2013). It's easy to infer that the delta pattern occurs while the subject is awake (processing cognitive tasks). Compared with baseline (a state with no delta waves), delta waves are associated with an increase in activity in many brain regions, which suggests that the delta pattern is not associated with a state of quiescence, but rather associated with an active state during which brain activity is consistently synchronized to the slow oscillation in specific cerebral regions (Dang-Vu *et al.*, 2008). Moreover, there is evidence that the delta pattern is primarily created in the hypothalamus (McGinty *et al.*, 1994), which is associated with a series of life-sustaining bodily functions such as autonomic regulation (e.g., blood pressure, heart rate, thermoregulation) and neuroendocrine control (Yoshida *et al.*, 2009). Considerable evidence supporting the association between delta waves and autonomic and metabolic processes shows that

Table 9.2: The intersubject correlation coefficients. "S" denotes subject. "Full" denotes the undecomposed full-frequency band data. The lower coefficients indicate that the subject's EEG data is more easily distinguished.

Subject		S 1	S 2	S 3	S 4	S 5	S 6	S 7	S 8	STD	Average
	Delta	0.137	0.428	0.246	0.179	0.221	0.119	0.187	0.239	0.089554	**0.219**
	Theta	0.447	0.671	0.552	0.31	0.387	0.207	0.199	0.386	0.151929	0.395
Patterns	Alpha	0.387	0.629	0.615	0.377	0.299	0.306	0.283	0.457	0.128653	0.419
	Beta	0.249	0.487	0.329	0.308	0.281	0.307	0.238	0.441	0.083224	0.33
	Gamma	0.528	0.692	0.538	0.362	0.521	0.667	0.428	0.537	0.102288	0.534
	Full	0.333	0.329	0.408	0.304	0.297	0.621	0.302	0.447	0.104231	0.38

the integration of cerebral activity with homeostatic processes might be one of the delta wave's functions (Knyazev, 2012). Since the life-sustaining functions are operational all the time, we can argue that, regardless of the state of the individual, delta oscillations will always be produced.

Next, we present some qualitative arguments to demonstrate that delta patterns contain the most distinctive information. We analyze intersubject correlations of the decomposed EEG patterns, which measure the similarity of two samples belonging to different subjects. For example, the intersubject correlation of subject 1 is calculated by the following steps: (1) randomly select 10 samples from subject 1; (2) randomly select 10 samples from each of other subjects (subjects 2–8) to obtain 70 samples; (3) calculate the pairwise similarity between the first 10 samples and the latter 70 samples to obtain 700 similarities; and (4) average the 700 similarities to produce the final intersubject correlation coefficient of subject 1. We measure the intersubject correlations for all of the frequency patterns in order to discover the most effective pattern. The correlation coefficient analysis results are shown in Table 9.2. We can observe that the delta pattern has the lowest intersubject correlation coefficients compared with the other patterns. This indicates that the delta patterns are most dissimilar to the other samples, and thus the most distinctive. Therefore, delta patterns show the most promise for user identification.

9.1.3 *Methodology*

9.1.3.1 *Overview*

Figure 9.2 outlines the specific steps for the proposed MindID system. The brain waves are collected by the portable EEG acquisition equipment while the user is in a relaxed state with his/her eyes closed (our preliminary experiment results illustrate that the delta wave is more dominant in

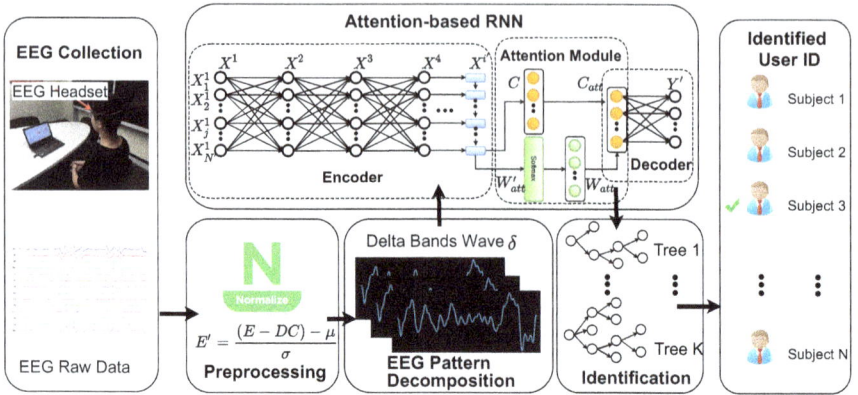

Fig. 9.2: Person identification system flowchart. At the beginning of identification, raw EEG data E is collected from the user and then fed to the preprocessing stage. The preprocessed data E' is decomposed to delta pattern δ, which serves as the input into the attention-based RNN. The encoder compresses the input sequence X^1 into an intermediate code C and simultaneously produces the weights W'_{att}. The attention-based module accepts both C and W'_{att} from the LSTM layer $X^{i'}$, processes W'_{att} through a softmax layer, and calculates the attention-based code C_{att}. Finally, a statistical boosting classifier is employed to identify the user.

the relaxed state, although it persists in all states). Each EEG sample is a numerical feature vector with N dimensions which correspond to the number of channels of the wearable EEG headset. The EEG samples are first preprocessed to remove the direct current (DC) offset, which is followed by normalization (Section 9.1.3.2). Next, we employ EEG pattern decomposition to isolate the delta waves, since they contain the most distinctive information that can be used to identify the subject (as outlined in Section 9.1.2). The delta waves are fed to an attention-based Encoder–Decoder RNN, which identifies the most distinctive channels and adjusts the weights accordingly. This model learns the deep correlations between the delta patterns which are then fed to a statistical boosting classifier (Section 9.1.3.5) to identify individual users.

9.1.3.2 *Preprocessing*

The raw EEG samples are preprocessed to remove the DC offset and normalize the signals. Eliminating the DC offset is necessary because EEG headsets invariably introduce a constant noise component into the recorded signals. In the preprocessing stage, this constant DC offset is first subtracted from the raw signal E. Our experiments (not shown for brevity) indicated that z-score scaling is the most suited to EEG data. The preprocessed data E' can be calculated by

$$E' = \frac{(E - DC) - \mu}{\sigma} \qquad (9.1)$$

where DC denotes the direct current which is 4200 muV,[3] μ denotes the mean of E-DC and σ denotes the standard deviation.

9.1.3.3 *EEG Pattern Decomposition*

In Section 9.1.2, we used empirical EEG data to show that the part of the EEG signals that belongs to the delta frequency band (0.5–3.5 Hz) is particularly well suited for accurate and robust user identification. To isolate the signals in the delta band, we use a Butterworth band-pass filter of order 3 with a frequency range of 0.5–3.5 Hz. The designed filter has the following specifications: the order is three, the low cut is 0.5 Hz, and the high cut is set as 3.5 Hz. The preprocessed signal Et is fed as input to this filter, which produces the decomposed delta pattern δ as output.

9.1.3.4 *Attention-Based RNN*

Next, the delta pattern δ is fed into an attention-based RNN structure (Wang *et al.*, 2016c) which aims to learn the most representable features for user identification. The general Encoder–Decoder RNN framework assumes that all feature dimensions of the input sequence are equally important and assigns them equal weight. In the context of EEG data, each dimension refers to a different electrode of the EEG equipment. For example, the first dimension (first channel) collects the EEG data from the $AF3$[4] electrode, which is located at the frontal lobe of the scalp, while the seventh dimension is gathered from $O1$ electrode at the occipital lobe.

Since different EEG channels record different aspects of the brain signals, some of which may be more representative of the individual, an approach that assumes all dimensions to be equal may not be optimal. In other words, the various EEG channels make unequal contributions to the person identification task and should be assigned weights accordingly.

---

[3] https://www.bci2000.org/mediawiki/index.php/Contributions:Emotiv. Accessed Dec. 15, 2020.
[4] Both $AF3$ and $O2$ are EEG measurement positions in the international 10–20 system.

The effectiveness of attention-based RNN has been demonstrated in various domains, including wearable sensor-based activity recognition (Chen *et al.*, 2018; Zhang *et al.*, 2018f), natural language processing (Ba *et al.*, 2014; Li *et al.*, 2017b; Wang *et al.*, 2016a), computer vision (Luong *et al.*, 2015) and speech recognition (Bahdanau *et al.*, 2016; Chan *et al.*, 2017). Encouraged by the wide success of this approach, we introduced the attention mechanism to the Encoder–Decoder RNN model to assign varying weights to specific dimensions of the EEG data. The proposed attention-based Encoder–Decoder RNN consists of three components (as shown in Figure 9.2): the encoder, the attention module, and the decoder. The encoder is designed to compress the input delta δ wave into a single intermediate code C; the attention module calculates a better intermediate code C_{att} by generating a sequence of distinct weights W_{att} for the different dimensions; the decoder accepts the attention-based code C_{att} and decodes it to the user ID. Note, this user ID is predicted by the attention-based RNN instead of by MindID; the final identified ID of the MindID approach will be introduced in Section 9.1.3.5.

Suppose the data in the ith layer could be denoted by $X^i = \{X^i_j; i \in [1, 2, \ldots, I], j \in [1, 2, \ldots, N^i]\}$ where j denotes the jth dimension of X^i. I represents the number of neural network layers in the proposed attention-based RNN model, and N^i denotes the number of dimensions in X^i. Taking the first layer as an example, we have $X^1 = \delta$, which indicates that input sequence is the delta pattern. Let the output sequence be $Y = \{Y_k; k \in [1, 2, \ldots, K]\}$, where K denotes the number of user ID categories. In this section, the user ID is represented by the one-hot label with length K. For simplicity, let's define the operation $\mathcal{T}(\cdot)$ as:

$$\mathcal{T}(X^i) = X^i W + b \qquad (9.2)$$

Furthermore, we have

$$\mathcal{T}(X^{i-1}_j, X^i_{j-1}) = X^{i-1}_j * W' + X^i_{j-1} * W'' + b' \qquad (9.3)$$

where W, b, W', W'', b' denote the corresponding weights and biases parameters.

The encoder component contains several nonrecurrent fully connected neural network layers and one recurrent LSTM layer. The nonrecurrent layers are employed to construct and fit a nonlinear function to purify the input delta pattern, the necessity of which has been demonstrated by our preliminary experiments.[5] The data flow in these nonrecurrent layers is calculated as follows:

$$X^{i+1} = \mathcal{T}(X^i) \tag{9.4}$$

The LSTM layer is adopted to compress the output of nonrecurrent layers to a fixed length sequence which is regarded as the intermediate code C. Supposing LSTM is the i'th layer, the code equals to the output of LSTM, which is $C = X_j^{i'}$. The $X_j^{i'}$ can be measured by

$$X_j^{i'} = \mathcal{L}(c_{j-1}^{i'}, X_j^{i-1}, X_{j-1}^{i'}) \tag{9.5}$$

where $c_{j-1}^{i'}$ denotes the hidden state of the $(j-1)$th LSTM cell. The operation $\mathcal{L}(\cdot)$ denotes the calculation process of the LSTM structure, which can be inferred from Equations (3.8)–(3.13).

The attention module accepts the final hidden states as the unnormalized attention weights W'_{att}, which can be measured by the mapping operation $\mathcal{L}'(\cdot)$ (similar to Equation (9.5))

$$W'_{att} = \mathcal{L}'(c_{j-1}^{i'}, X_j^{i-1}, X_{j-1}^{i'}) \tag{9.6}$$

and calculate the normalized attention weights W_{att}

$$W_{att} = softmax(W'_{att}) \tag{9.7}$$

The softmax function is employed to normalize the attention weights into the range of $[0, 1]$. Therefore, the weights can be explained as the probability that the code C is relevant to the output results. Under the attention mechanism, the code C is weighted to C_{att}

$$C_{att} = C \odot W_{att} \tag{9.8}$$

Note, C and W_{att} are trained instantaneously. The decoder receives the attention-based code C_{att} and decodes it to predict the user's identity Y'.[6]

---

[5]Some optimal designs, like the neural network layers, are validated by the preliminary experiments, but the validation procedure is not reported in this section for reasons of space.

[6]Note: Y' is not the identification result of the MindID model. The final identified user ID is I_D, calculated in Equation (9.11).

Since Y' is predicted at the output layer of the attention-based RNN model $(Y' = X^I)$, we have

$$Y' = \mathcal{T}(C_{att}) \tag{9.9}$$

Finally, we employ the cross-entropy function to calculate the prediction cost between the predicted ID Y' and the ground truth Y. ℓ_2-norm (with parameter λ) is selected to prevent overfitting. The cost is optimized by the Adam optimizer algorithm (Kingma and Ba, 2014). The threshold for the number of iterations of the attention-based RNN is set as n_{iter}. The weighted code C_{att} has a linear relationship with the output layer and the predicted results. If the model is well trained, then the weighted code C_{att} could be regarded as a high-quality representation of the identity of the user. We set the learned deep feature X_D equal to C_{att}, $X_D = C_{att}$, and employ it to recognize the user in the identification stage.

9.1.3.5 *Identification*

In this section, we employ the eXtreme Gradient Boosting Classifier (XGBoost) (Chen and Guestrin, 2016) to classify the learned deep feature X_D for user identification. The XGBoost Classifier fuses a set of classification and regression trees (CART) and exploits information as detailed as possible from the input features X_D. It builds multiple trees, each of which has its leaves and corresponding scores. Moreover, it proposes a regularized model formalization to prevent overfitting and is widely used for its accurate prediction power.

The learned deep feature X_D is used to train a number of the CART (there are M trees) and predict a set of user IDs. Suppose $x_d \in X_D$ is a single sample of the deep feature. The final identification result of the input x_d is calculated as

$$y_m = f(x_d) \tag{9.10}$$

$$I_D = F\left(\sum_{1}^{M} y_m\right), \quad m = 1, 2, \ldots, M \tag{9.11}$$

where f denotes the classification function of a single tree, y_m denotes the predicted ID of the mth tree and F denotes the mapping from the single tree prediction space to the final prediction space. The I_D is the final identified user ID. The overall procedure of identification is summarized in Algorithm 4. (For further details — e.g., experiments, latency, band comparison — please refer to Zhang *et al.* (2018g).)

Algorithm 4 The User Identification Algorithm

Input: EEG raw data E

Output: Identification results I_D

1: Initialization

2: Preprocessing: $E' \leftarrow E$

3: EEG pattern decomposition: $\delta \leftarrow E'$

4: **if** $iteration < n_{iter}$ **then**

5:      **for** $i = 1, 2, \ldots, I$ **do**

6:        $X^1 = \delta$

7:        $C \leftarrow X^1, \mathcal{L}(c_{j-1}^{i'}, X_j^{i-1}, X_{j-1}^{i'})$

8:        $W_{att} \leftarrow C, \mathcal{L}'(c_{j-1}^{i'}, X_j^{i-1}, X_{j-1}^{i'})$

9:        $C_{att} = C \odot W_{att}$

10:       $X_D = C_{att}$

11:      **end for**

12: **else**

13:      Return X_D

14: **end if**

15: **for** X_D **do**

16:      $I_D \leftarrow X_D$

17: **end for**

18: **return** I_D

9.1.4 *Discussions*

Here, we discuss the challenges facing the EEG-based person identification system.

First, EEG-based identification is less vulnerable to attacks compared to existing biometric identification systems. In order to evaluate the attack resilience of MindID, we tested our approach to dealing with the threat posed by unauthorized subjects from a number of attack categories (Pasqualetti *et al.*, 2013). We randomly selected 10 subjects from the EEG-S-L data set as authorized users, while the rest of the users were defined as unauthorized subjects. During testing, given the unauthorized users' EEG signals, MindID calculated the probability that each sample belonged to an authorized subject. A specific user was regarded as unauthorized if the prediction probability was below the minimum threshold. Our experimental results demonstrate that MindID is able to precisely detect the authorized users with around 99% accuracy given an appropriate threshold setting. This suggests that our approach also has the potential to distinguish an attack from an unauthorized subject.

EEG signals are known to be sensitive to various factors, such as the mood of the subject or their intake of foods, drugs, or alcohol. Knyazev (2012) infers that EEG signals are also affected by inherent factors, such as panic, sustained pain, sexual arousal, and so on. Dubbelink *et al.* (2008) conducted experiments on obese and lean female adolescents and recorded the MEG signal of participants' brains. The obese adolescents had increased synchronization in delta and beta frequency bands compared to the lean controls. Reid *et al.* (2006) claim that the increase of delta power during the first five minutes following use of cocaine correlated with increased ratings of cocaine craving. Reward-related decrease of delta activity has been observed after administration of legal psychoactive drugs, such as alcohol (Sanz-Martin *et al.*, 2011), tobacco (Knott *et al.*, 2008), and caffeine (Hammond, 2003). One scope of our future work will be to study how the identification system is influenced by the aforementioned factors in order to enhance the adaptability of the current approach.

The impact of population size on performance needs further investigation. However, in this section, our results already demonstrate that MindID could be used in settings such as small offices accessed only by a small group of people.

The EEG data of an individual is known to change gradually with environmental factors such as age, mental state and lifestyle. For example, delta patterns are known to decrease with age (Emek-Savaş *et al.*, 2016). This suggests that the pretrained model used in MindID should be updated when such changes are detected.

While we provide some explanations in Section 9.1.2 as to why delta patterns may be the most informative for user identification, the underlying mechanism is still not well understood. Further investigation is necessary.

9.2   Person Authentication

Based on the EEG-based identification system introduced in the previous section, we here propose a dual-authentication system via brain waves and gait signals for higher reliability. The person identification task aims only to confirm the user's identity if the user's profile is prestored in the data set, whereas the authentication attempts to detect whether the user is authenticated (i.e., whether the profile of the requester matches with the profile prestored in the data set).

9.2.1 *Motivations*

The advantages of the EEG-based system are introduced in the last section. On the other hand, gait-based authentication has been an active topic for years (Qian *et al.*, 2018; Yao *et al.*, 2018). Gait data are more generic and are easily gathered from popular inertial sensors. Gait data are also unique because they are determined by intrinsic factors (e.g., gender, height, and limb length), temporal factors (Callisaya *et al.*, 2009) (e.g., step length, walking speed, and cycle time), and kinematic factors (e.g., joint rotation of the hip, knee, and ankle, mean joint angles of the hip/knee/ankle, and thigh/trunk/foot angles). In addition, a person's gait behavior is established inherently over the long term and therefore difficult to fake. Hoang and Choi (2014) have proposed a gait-based biometric authentication system to analyze gait data gathered by mobile devices, adopted error correcting codes to process the variation in gait measurement, and finally achieve a false acceptance rate (FAR) of 3.92% and a false rejection rate (FRR) of 11.76%. Cola *et al.* (2016) collected wrist signals and trained gait patterns to detect invalid subjects (unauthenticated people); their proposed method achieved an equal error rate (EER) of 2.9%.

Despite these tremendous efforts, various challenges still face single EEG-/gait-based authentication systems: (1) the solo EEG-based authentication system is highly fake-resistant and has excellent authentication performance but is easily affected by environmental factors (e.g., noise), subjective factors (e.g., mental state), and noisy brain signals (e.g., not concentrating); (2) the solo gait-based authentication system is more stable across different scenarios but has relatively low performance; (3) the solo EEG/gait authentication system generally obtains a FAR (which is extremely crucial in highly confidential authentication scenarios)[7] higher than 3% (Hoang and Choi, 2014), meaning it is not precise enough for highly sensitive places such as military bases, the treasuries of banks, and political offices, where tiny misjudgements could result in economic or political catastrophes; and (4) the single authentication system may break down under attack, but it provides for no backup plan.

In this section, we propose DeepKey, a novel biometric authentication system that enables dual authentication, leveraging the advantages of both gait-based and EEG-based systems. Compared with either the gait-based

---

[7]In highly confidential authentication scenarios, FAR is more crucial than other metrics, such as accuracy.

or EEG-based authentication system, a dual authentication system offers more reliable and precise identification. DeepKey consists of three main components: the Invalid ID Filter Model to eliminate invalid subjects, the EEG Identification Model to identify EEG IDs, and the Gait Identification Model to identify gait IDs. An individual is granted access only after s/he passes all the authentication components.

To the best of our knowledge, it is a very advanced idea to combine EEG and gaits for identity authentication. Taking advantage of both EEG and gait signals, this combination is expected to significantly improve authentication reliability. We emphasize several differences compared to MindID (Section 9.1): (1) this model includes an invalid ID filter, whereas MindID only focuses on identification; (2) this model adopts two biometrics, EEG and gait, whereas MindID uses only EEG signals; and (3) this model conducts extensive real-world experiments to collect gait signals.

We also present a high-level comparison of different biometric authentication systems (Table 9.3). In an authentication system, the FAR should take higher priority over other metrics (i.e., accuracy and FRR) as even a tiny misjudgment (false positive) may engender catastrophic consequences.

9.2.2 *Methodology*

9.2.2.1 *Overview*

The DeepKey system is intended for deployment to oversee access to confidential locations (e.g., bank vault), military bases, and government confidential residences). As shown in Figure 9.3, the overall workflow of the DeepKey authentication system consists of the following four steps:

(1) Step 1: EEG data collection. The subject requesting authentication is required to wear the EEG headset and remain relaxed. The collection of EEG data ($\mathcal{E}$) will typically take two seconds.
(2) Step 2: Gait data collection. The subject takes off the EEG headset and puts on three inertial measurement units (IMUs), then walks through an aisle to allow collection of gait data $\mathcal{G}$ by the IMUs.
(3) Step 3: DeepKey authentication. The gathered EEG and gait data are flattened and associated with input data $\mathcal{I} = [\mathcal{E} : \mathcal{G}]$ to be fed into the DeepKey authentication algorithm.
(4) Step 4: Decision. The decision to approve or deny will depend on the DeepKey authentication results.

Table 9.3: System-level comparison between DeepKey and other biometric authentication systems.

	Reference	Biometric	Method	#-Subject	Data Set	Accuracy	FAR	FRR
Uni-modal	Cola et al. (2016)	Gait	Semi-ANN	15	Local	97.4		
	Muramatsu et al. (2015)	Gait	AVTM-PdVs	100	Public	77.72		
	Al-Naffakh et al. (2017)	Gait	MLP	60	Local	99.01		
	Sun et al. (2014)	Gait	Voting classifier	10	Local	98.75		
	Konno et al. (2015)	Gait	Two SVMs	50	Local		1.0	1.0
	Thomas and Vinod (2017)	EEG	PSD	109	Public		1.96	1.96
	Chuang et al. (2013)	EEG	Threshold	15	Local		0	2.2
	Gui et al. (2014)	EEG	Wavelets + ANN	32	Local	90.03		
	Bashar et al. (2016)	EEG	BFIR + SVM	9	Local	94.44		
	Thomas and Vinod (2016b)	EEG	IAF + Delta EEG	109	Public	90.21		
	Jayarathne et al. (2016)	EEG	CSP + LDA	12	Local	96.97		
	Zhang et al. (2018g)	EEG	RNN + XGB	8	Local	98.82		
	Keshishzadeh et al. (2016)	EEG	AR + SVM	104	Public	97.43		
	Long et al. (2012)	Fingerprint	ZM + ANN	40	Public	92.89	7.108	7.151
		Face					11.52	13.47
		Fusion					4.95	1.12
	Yano et al. (2012)	Iris	Wavelets	59	Public		13.88	13.88
		Pupil	+ Hamming				5.47	5.47
		Fusion					2.44	2.44
Multi-modal	Manjunathswamy et al. (2015)	ECG	Wavelet	50	Public		2.37	9.52
		Fingerprint	Histogram				7.77	5.55
		Fusion	Score fusion				2.5	0
	Derawi and Voitenko (2014)	ECG	Cross correlation	30	Local		4.2	4.2
		Gait					7.5	7.5
		Fusion	Score fusion				1.26	1.26
	Ours	EEG	Delta EEG +	7	Local	99.96		
		Gait	Att-RNN + KNN			99.61		
		Fusion				**99.57**	**0**	**1.0**

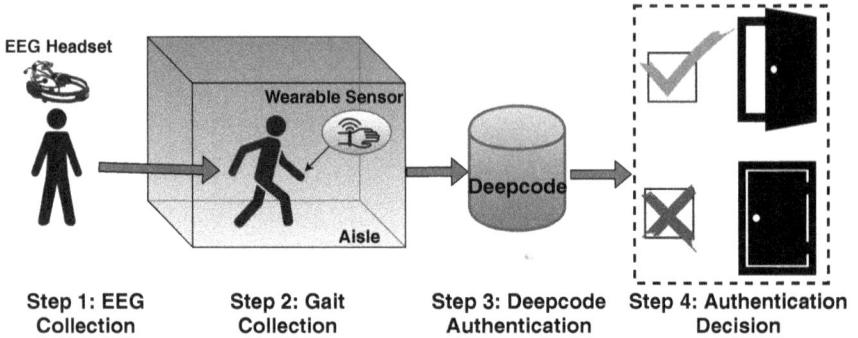

Step 1: EEG Collection **Step 2: Gait Collection** **Step 3: Deepcode Authentication** **Step 4: Authentication Decision**

Fig. 9.3: The DeepKey authentication system workflow. The EEG and gait data collection are cascading.

The third step is the most crucial, where the DeepKey authentication system receives the associated input data $\mathcal{I}$ and accomplishes two goals: authentication and identification. In terms of authentication, we aim to find user's identity based on EEG signals as EEG is highly fake-resistant. EEG signals are invisible and unique, making it difficult to be duplicated or hacked. For the latter goal, we adopt a deep learning model to extract the distinctive features and feed them into a nonparametric neighbor-based classifier for ID identification. In summary, the DeepKey authentication algorithm contains several key stages, namely *Invalid ID Filter*, *Gait-Based Identification*, *EEG-Based Identification* and *Decision Making*. The overall authentication contains the following several stages (Figure 9.4; data flow is presented in Algorithm 5):

(1) Based on the EEG data, the Invalid ID Filter either classifies the subject as genuine or an impostor. If the subject is classed as an impostor, the request will be denied.
(2) If the individual is determined to be genuine, the EEG/Gait Identification Model will confirm the individual's authorized EEG/Gait ID. This model is pretrained offline with the attention-based LSTM model (Section 9.2.2.3). The output is the ID number associated with the person's detailed personal information.
(3) The final stage is to check for consistency between the EEG ID and the Gait ID. If they are identical, the system will grant an approval; otherwise, the subject will be denied access and corresponding security measures will be taken.

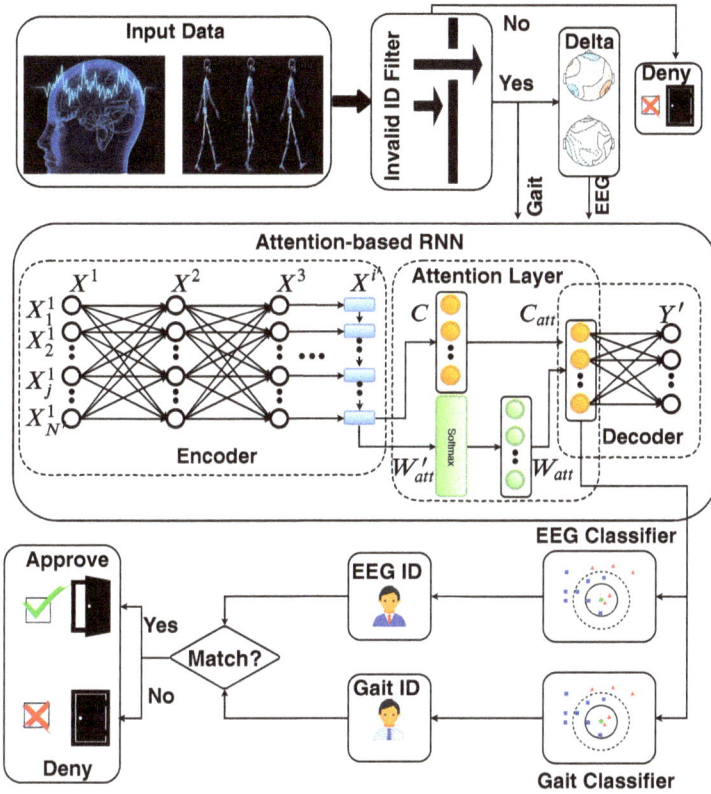

Fig. 9.4: Authentication workflow. If the input data cannot pass the invalid ID filter, it is immediately regarded as an impostor and denied access. If passed, the delta pattern and gait signals are fed into an attention-based RNN structure in parallel to study the distinctive features C_{att}. The learned features are classified by the EEG and Gait classifiers in order to identify the subject's EEG and Gait IDs. The subject is approved only if the EEG ID is a match with the Gait ID.

9.2.2.2 *Invalid ID Filter Model*

Through our preliminary experiments, we have found that raw EEG signals, compared with raw gait data, have better characteristics for preventing invalid ID due to the high fake-resistance of EEG data and the richness of distinguishable features in EEG signals. Thus, in this section, we use only EEG data for invalid ID filtering.

Subjects in an authentication system are categorized into two classes: *authorized* and *unauthorized*. Since unauthorized data are not available in the training stage, an unsupervised learning algorithm is required to identify an invalid ID. In this section, we apply one-class SVM to sort

---

Algorithm 5 DeepKey System

Input: EEG data $\mathcal{E}$ and Gait data $\mathcal{G}$
Output: Authentication Decision: Approve/deny
1: #Invalid ID Filter:
2: **for** $\mathcal{E}, \mathcal{G}$ **do**
3:    Genuine/Impostor $\leftarrow \mathcal{E}$
4:    **if** Impostor **then**
5:      **return** Deny
6:    **else if** Genuine **then**
7:      #EEG Identification Model:
8:      **while** $iteration < n_{iter}^E$ **do**
9:        $X^{i+1} = tanh(\mathcal{T}(X^i)) \; \{X^1 = \mathcal{E}\}$
10:        $C = X_j^{i'} = \mathcal{L}(c_{j-1}^{i'}, X_j^{i-1}, X_{j-1}^{i'})$
11:        $W_{att} = softmax(\mathcal{L}'(c_{j-1}^{i'}, X_j^{i-1}, X_{j-1}^{i'}))$
12:        $C_{att} = C \odot W_{att}$
13:        $E_{ID} \leftarrow C_{att}$
14:      **end while**
15:      #Gait Identification Model:
16:      **while** $iteration < n_{iter}^G$ **do**
17:        $G_{ID} \leftarrow \mathcal{G}$
18:      **end while**
19:      **if** $E_{ID} = G_{ID}$ **then**
20:        **return** Approve
21:      **else**
22:        **return** Deny
23:      **end if**
24:    **end if**
25: **end for**

---

out the unauthorized subjects. Given a set of authorized subjects $\mathcal{S} = \{S_i, i = 1, 2, \ldots, L^\circ\}$, $S_i \in R^{n_s}$, where L° denotes the number of authorized subjects and n_s denotes the number of dimensions of the input data, the input data consists of EEG data $\mathcal{E} = \{E_i, i = 1, 2, \ldots, L^\circ\}$, $E_i \in R^{n_e}$ and gait data $\mathcal{G} = \{G_i, i = 1, 2, \ldots, L^\circ\}$, $G_i \in R^{n_g}$. n_g and n_e denote the number of dimensions of the gait data and EEG data, respectively, and $n_s = n_g + n_e$.

For each authentication, the collected EEG data E_i includes a number of samples. Each sample is a vector with shape (1, 14), where 14 denotes the number of electric nodes in the Emotiv headset. To trade off on the authentication efficiency (for less collection and waiting time) and computational performance, based on experimental experience, we fed 200 samples (200, 14) into the Invalid ID Filter. The 200 EEG samples we collected in 1.56 s indicate that the signal-collecting latency is acceptable. The final filter result is the mean of the results for all the samples.

9.2.2.3 *EEG Identification Model*

Compared to gait data, EEG data contain more noise and are thus more challenging to handle. Given the complexity of EEG signals, data preprocessing is necessary. In practical EEG data analysis, the assembled EEG signals can be divided into several different frequency patterns (delta, theta, alpha, beta, and gamma) based on the strong intra-band correlation with a distinct behavioral state. The EEG frequency patterns and the corresponding characters are listed in Table 2.2 (Zhang *et al.*, 2018g). The previous chapter has demonstrated by qualitative analysis and experimental results that the delta pattern, compared with other EEG patterns, contains the most distinctive information and is the most stable pattern in different environments. Thus, we adopt a bandpass Butterworth filter (0.5 Hz to 3.5 Hz) to extract delta wave signals for further authentication. For simplicity, we denote the filtered EEG data as $\mathcal{E}$.

This calculation procedure is very similar to the attention-based RNN in Section 9.1.3.4. We directly jump to the output of the identification model. Set the learned deep feature X_D to C_{att}, $X_D = C_{att}$, and feed it into a lightweight nearest neighbor classifier. The EEG ID, which is denoted by E_{ID}, can be directly predicted by the classifier.

The Gait Identification Model works similarly to the EEG Identification Model except for the frequency band filtering. The iteration threshold of attention-based RNN is set as n_{iter}^G. The Gait Identification Model receives subjects' gait data $\mathcal{G}$ from the input data $\mathcal{I}$ and maps to the user's Gait ID G_{ID}. All the model structures, hyperparameters, optimization, and other settings in the EEG and Gait Identification Models remain the same to retain the lower complexity of the DeepKey system.

9.2.3 **Data Acquisition**

We designed real-world experiments to collect cascading EEG data and gait data.[8] The experiments (Figure 9.5) were conducted on seven healthy

---

[8]This experiment was approved by our ethics board.

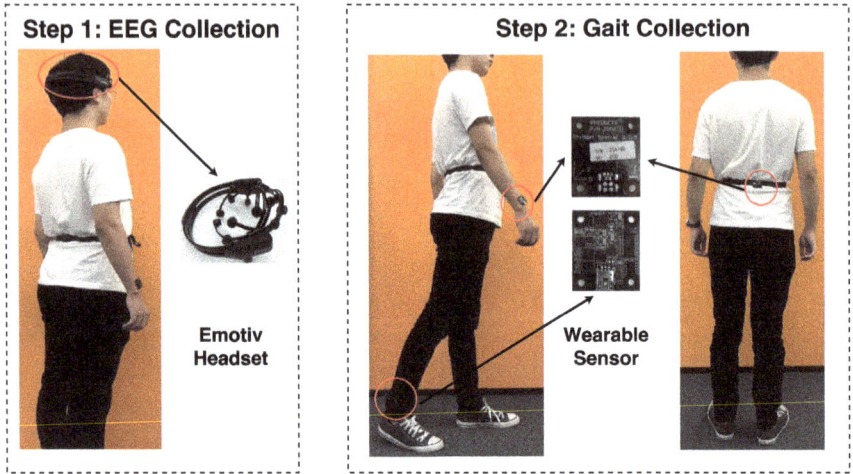

Fig. 9.5: Data collection. The two collection steps are cascading to eliminate the impact on EEG data of walking. Solo EEG signals are collected in the first step, the gait signals in the second.

participants, four males and three females, aged 26 ± 2. In Step 1, each participant remained standing and relaxed with eyes closed. The EEG data were collected by EPOC+ Emotiv headset[9] which integrated 14 electrodes (corresponding to 14 EEG channels) with a sampling rate of 128 Hz. In Step 2, each participant walked through an aisle to generate the gait data. In the gait collection procedure, three IMUs were attached — one to the participant's left wrist, another to the middle of the back, and a third to the left ankle. Each IMU (PhidgetSpatial 3/3/3[10]) used an 80 Hz sampling rate to gather nine dimensional motor features, contained a three-axis accelerometer, three-axis gyroscope, and three-axis magnetometer.

[9]https://www.emotiv.com/product/emotiv-epoc-14-channel-mobile-eeg/. Accessed Jul. 7, 2019.
[10]https://www.phidgets.com/?&prodid=48. Accessed Dec. 15, 2020.

Chapter 10

Visual Reconstruction

In this chapter, we investigate the use of deep learning algorithms to reconstruct the visual stimulus by decoding the user's brain signals.

10.1 Brain2Object: Printing Your Mind

In this section, we propose a unified approach to learn the robust structured EEG feature representations for recognizing the imagery of objects seen by the individual. We first design a multiclass Common Spatial Pattern (CSP) for distilling the compact representations. CSP has proven successful in extracting features using eigendecomposition based on the variance ratio between different classes (Elisha *et al.*, 2017). Next, we propose dynamic graph representation (DGR) of EEG signals to adaptively embed the spatial relationship among the channels (each channel represents one EEG electrode) and their neighbors by learning a dynamic adjacency matrix. Finally, a CNN is employed to aggregate higher-level spatial variations from the transformed graph representations.

Built on top of the aforementioned computational framework, we present a mind-controlled end-to-end system with an integrated graphic interface called Brain2Object. It enables an individual to print a physical replica of an object that s/he is observing by interpreting visually evoked EEG signals in a real time manner. To enable the end-to-end workflow, the proposed system gathers the user's brain activities through EEG acquisition equipment and forwards the collected EEG data to a pretrained model, which automatically recognizes the object that the user is observing. Imagine that a child observes a toy, for example, Pinkie Pie (from *My Little Pony*) belonging to her friend and likes it very much and wishes that she could have one, too. Brain2Object can make that wish a reality by translating her brain signals into commands to the 3D printer to fabricate a copy. The ability to print a replica of any observed object could be of tremendous value to a variety of professionals, including engineers, artists, construction workers, students, teachers, law enforcement, urban planners, and so on. By automating the process, Brain2Object takes the mystery of reading the human mind out of the realm of the experts and opens up the possibility of a wide range of brain signal applications useful to the masses.

To summarize, we present an end-to-end digital fabrication system, Brain2Object, in addition to the precise decoding of human brain signals that together allow an individual to instantly create a real-world replica (or model) of any object within view. Furthermore, a prototype implementation demonstrates the practicality of Brain2Object.

10.1.1 *Brain2Object System*

The overall aim of Brain2Object is to automatically recognize the object that the user wishes to fabricate by analyzing her visually evoked brain signals and actuating a 3D printer accordingly. As shown in Figure 10.1, a brain signal recognition model employs multiclass CSP to separate the multivariate signals into additive subcomponents which have maximum differences. The spatial dependencies among the processed data are extracted by DGR and then forwarded to the CNN for recognition. CSP is widely used in brain research to find spatial filters which can maximize variance between classes and achieves competitive performance (Meisheri *et al.*, 2018). In this section, we adopt the one-versus-others strategy for multiclass CSP analysis.

10.1.1.1 *System Overview*

This Brain2Object system aims to generate a physical replica based on user's mental activity. As shown in Figure 10.1, the system is made up of an offline and online component.

The offline component is aimed at building a robust and effective unified classification model capable of recognizing the specific object observed by the user by analyzing the corresponding brain signals evoked during this process. For this purpose, we first record the EEG signals of individuals while they are observing a wide range of objects (the details will be

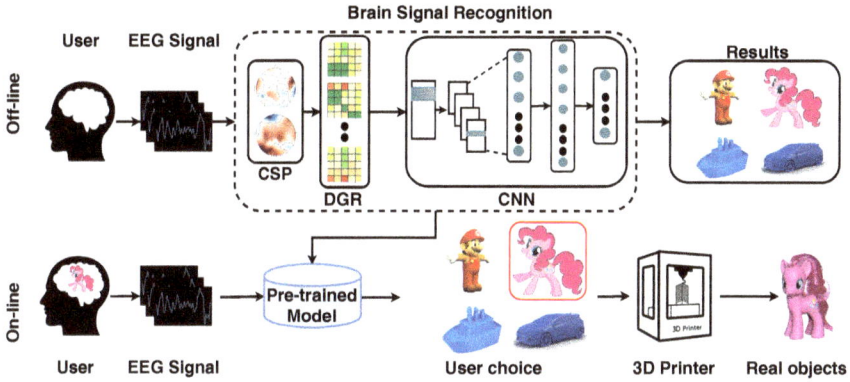

Fig. 10.1: Overview of Brain2Object. The object (e.g., Pinkie Pie) observed by the user is reflected in the visually evoked EEG signals, which can be accurately recognized by the pretrained recognition model. The recognition module employs multiclass CSP to separate the multivariate signals into additive subcomponents which have maximum differences. The spatial dependencies among the processed data are extracted by DGR and then forwarded to the CNN for recognition. The schematic of the identified object is loaded from the model library of the 3D printer to fabricate a replica. This algorithm considers both graphical and spatial information from EEG signals.

introduced in Section 10.1.2). Next, the gathered EEG data are analyzed using multiclass CSP (Chin *et al.*, 2009) to extract the eigenvalues of various categories of objects. CSP has been widely used in EEG signal analysis, such as classification of motor imagery EEG signals (Kang and Choi, 2014; Wang *et al.*, 2006), and achieves competitive performance. Thus, in our system, we adopt CSP for discriminative spatial filtering to enhance the SNR of EEG signals. Considering the global spatial relationship among EEG channels, we propose DGR to transform the CSP-processed signals to a new space since graph representation has been shown to be helpful in refining and capturing spatial information (Song *et al.*, 2018). In the embedded space which encompasses the topological structure among functional brain lobes, each channel not only represents the amplitude of the measured signals but also reflects the interdependence with other channels. CNNs are widely used in the processing of 2D data in applications such as image recognition (Ciresan *et al.*, 2011), human activity classification (Ning *et al.*, 2018), and object searching (Ren *et al.*, 2017), due to their salient features such as regularized structure, good spatial locality, and translation invariance. Thus, we employ CNN as a classifier to recognize the graphical features. After a number of training epochs, the converged classification model with stable weights is stored for online recognition.

During online operation, the user wears an EEG signal acquisition headset, while she concentrates on a physical object, which collects her brain signals in real time. The gathered signals are forwarded to the pretrained model, which attempts to recognize the object. For example, as shown in Figure 10.1, the user is focusing specifically on "Pinkie Pie" rather than any other ponies, and the pretrained model is trained to automatically recognize this object and send an appropriate command to the 3D printer, which loads the 3D physical model and fabricates a copy.

10.1.1.2 *Multiclass Common Spatial Filtering*

CSP is widely used in the BCI field to find spatial filters, which can maximize variance between the classes on which it has been conditioned (Meisheri *et al.*, 2018). It has been successfully used for recognition of movement-related EEG (Ramoser *et al.*, 2000). CSP was first introduced in binary classification problems but has since been extended to multiclass scenarios (Ang *et al.*, 2012). In this section, we adopt the one-versus-others strategy for multiclass CSP analysis. Assume the gathered EEG data can be denoted by $\mathbb{E} = \{\boldsymbol{E}_i, i \in 1, 2, \ldots, N\}$ where N denotes the number of samples and each sample by $\boldsymbol{E}_i \in \mathbb{R}^{M \times L}$ where M denotes the number of EEG channels and L denotes the number of time slices. For example, assume the EEG collection equipment contains 64 channels and has a sampling frequency of 260 Hz, then data collected for 1 s can be represented by a 2D matrix $\boldsymbol{E}_i \in \mathbb{R}^{64 \times 260}$, where each row denotes one channel and each column denotes one time slice. For each specific sample, we first calculate the covariance matrix as (Meisheri *et al.*, 2018)

$$C_i = \frac{\boldsymbol{E}_i \boldsymbol{E}_i^T}{trace(\boldsymbol{E}_i \boldsymbol{E}_i^T)} \tag{10.1}$$

where ^{T} refers to the transpose operation and $trace()$ denotes the sum of the diagonal values in $\boldsymbol{E}_i$. The covariance matrix $\boldsymbol{C}_i$ presents the correlation between different columns (sampling points) in $\boldsymbol{E}_i$. In other words, C_i captures the temporal dependencies in the EEG samples. Suppose there are overall $K \in \mathbb{R}$ categories corresponding to the total number of objects for which EEG data is collected. For each category, the average covariance matrix is computed as

$$\bar{C}_k = \frac{1}{N_k} \sum_{i=1}^{N_k} C_i \tag{10.2}$$

where N_k denotes the number of distinct samples of the kth category. The composite covariance matrix is defined by the sum of each category's covariance matrix as

$$\bar{C} = \sum_{k=1}^{K} \bar{C}_k \tag{10.3}$$

The eigenvalue λ and eigenvector U of $\bar{C}$ can be deduced as

$$\bar{C} = U\lambda U^T \tag{10.4}$$

In order to decorrelate the covariance matrix, we apply the whitening transformation to the eigenvector which is sorted descendingly by eigenvalues. The whitened matrix is calculated as

$$S = P\bar{C}P^T \tag{10.5}$$

where S has unit diagonal covariance and P can be represented by

$$P = \sqrt{\lambda^{-1}}U \tag{10.6}$$

Based on Equation (10.3), we have

$$S_i = \sum_{k=1}^{K} P\bar{C}_i P^T; \quad S = \sum_{k=1}^{K} S_i \tag{10.7}$$

Combining this with Equation (10.4), we obtain

$$S_i = \sum_{k=1}^{K} B\lambda_i B^T \tag{10.8}$$

where $B = PU$ can be regarded as the common eigenvector. Since S has unit diagonal covariance, we have

$$\sum_{k=1}^{K} \lambda_i = I \tag{10.9}$$

where I denotes the identity matrix.

The main purpose of CSP is to maximize the distance among various categories in a transformed space — in other words, to optimize the equation (Parra *et al.*, 2005)

$$\boldsymbol{w}^* = \arg\max \frac{\boldsymbol{w} \boldsymbol{S}_i \boldsymbol{w}^T}{\sum_{j \neq i} \boldsymbol{w} \boldsymbol{S}_j \boldsymbol{w}^T} \tag{10.10}$$

where $\boldsymbol{S}_j (j \neq i)$ denotes all other classes except class i (one-versus-others). Equation (10.10) is in the form of the Rayleigh quotient and the solution can be calculated by generalized eigenvalue

$$\boldsymbol{S}_i \boldsymbol{w} = \lambda \sum_{j \neq i} \boldsymbol{S}_j \boldsymbol{w} \tag{10.11}$$

The eigenvector of the above equation is the required transformation weights. All the EEG samples share the same weight $\boldsymbol{s}$. In case of the information loss, we employ the full eigenvectors. Generally, the EEG time series has more samples than the number of channels, i.e., $L > M$. Thus the transformation weights have the shape $[M, M]$ and the processed sample $\bar{\boldsymbol{E}} \in \mathbb{R}^{M \times L}$ is measured by $\bar{\boldsymbol{E}} = \boldsymbol{w}\boldsymbol{E}$.

10.1.1.3 *Dynamic Graph Representation*

We propose dynamic graph representation (DGR) to transform the CSP processed signals to a new space since graph representation has been shown to be helpful in refining and capturing spatial information (Song *et al.*, 2018). In post-CSP-processed EEG data $\bar{\boldsymbol{E}}$, each channel (row) separately provides the voltage amplitude of a specific electrode rather than the aggregated spatial information. The signals are discrete and discontinuous in the spatial domain. Hence, traditional spatial feature representation methods such as CNN are not well suited for further processing (Song *et al.*, 2018). Instead, we invoke the knowledge of the connections of the brain neurons to map $\bar{\boldsymbol{E}}$ to a new space where each element represents not only the specific channel amplitude but also the spatial relationship with its neighboring channels.

For this purpose, we regard the brain network as a complete weighted undirected graph with M vertices, where each vertex denotes a channel. The term "complete" denotes that each vertex is connected to every other vertex in this graph. The graph can be defined as $\mathcal{G} = \{\mathcal{V}, \mathcal{E}, \mathcal{A}\}$, where $\mathcal{V} \in \mathbb{R}^M$ denotes the set of vertices with the number of $|\mathcal{V}| = M$ and $\mathcal{E}$ denotes the set of edges connecting the vertices. Suppose $\mathcal{A} \in \mathbb{R}^{M \times M}$ denotes the adjacency matrix representing the connectivity within $\mathcal{V}$. In particular, the element in the ith row and jth column of the adjacency matrix measures the weight or importance of the edges between the ith and the jth vertices.

The graph representation is dynamic, which means that the elements of the adjacency matrix are adaptively updated with the evolution of the model during training, hence the name dynamic graph representation (DGR). Recall Figure 6.3 that illustrates an example of a complete weighted undirected graph which is the composite of five vertices reading from the frontal (F) and temporal (T) lobes of the human brain. The diagonal elements are zero since each vertex is not connected to itself. However, the proposed representation should also contain information representative of each individual vertex. To incorporate this information, we include an identity matrix I. The resulting DGR is

$$\boldsymbol{E}' = (\mathcal{A} + \boldsymbol{I})\bar{\boldsymbol{E}} \tag{10.12}$$

The represented data $\boldsymbol{E}'$ with shape $[M, L]$ can dynamically learn the intrinsic relationship between different EEG channels and thus benefit most from discriminative EEG feature extraction.

10.1.1.4 *Object Recognition*

The DGR representation of the EEG signals (with shape $[M, L]$) serves as input to a specified CNN structure for feature refining and classification. CNN could capture the distinctive dependencies among the patterns associated with different EEG categories. The designed CNN comprises one convolutional layer followed by three fully connected layers (as shown in Figure 10.1). (For details on convolutional layers, please refer to Section 3.1.3.) The softmax function is used for activation. For each EEG sample, the corresponding label information is presented by one-hot label $\boldsymbol{y} \in \mathbb{R}^K$. The error between the prediction and the ground truth is evaluated by cross-entropy loss following:

$$loss = -\sum_{k=1}^{K} \boldsymbol{y}_k log(p_k) \tag{10.13}$$

where p_k denotes the predicted probability of the observation of an object belonging to category k. The error is optimized by the Adam optimizer algorithm. A dropout layer is added after the first fully connected layer with a 0.5 dropout rate in case of overfitting.

10.1.2 *Data Acquisition*

We aimed to gather a local EEG data set reflecting the user's brain voltage fluctuations under visual stimulation by a number of object images. In the ideal environment, the system is expected to recognize the EEG pattern produced in response to a random image. However, as this is a first exploration of this idea, we limited our study to include images of only four objects: a car, a boat, Pinkie Pie Pony and Mario (from the video game).

We recruited eight healthy participants five males and three females aged 22–27 years, to participate in this study. The experiment was approved by our ethics committee (HC190315). Data collection was conducted in a quiet room. As shown in Figure 10.2, each subject wore the EPOC+ Emotiv EEG headset which contains 14 channels corresponding to the 10–20 system (an internationally recognized method used to apply and describe the location of scalp electrodes). The sampling rate was set at 128 Hz and the headset could wirelessly connect with the computer over Bluetooth. The participants sat in a comfortable armchair, maintained a relaxed state and gazed at a monitor placed approximately 0.5 m away. Each subject participated in 10 sessions and each session contained four trials.

Each trial lasted 15 s and was comprised of three phases, each lasting 5 s. In the first phase, the monitor showed an empty slide and the subject

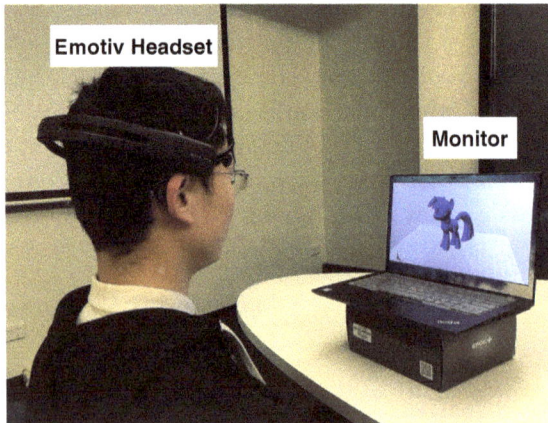

Fig. 10.2: Data acquisition experiment. The participant wears the EPOC+ Emotiv headset with 14 channels siting in front of a monitor which shows the Pinkie Pie.

was asked to relax. In the second phase, a picture of a random object was presented in the center of the screen and the subject was instructed to focus on the image. The final phase was identical to the first phase. Naturally, only EEG signals collected during the second phase were used in our data set. In the second phase, the image was chosen with equal probability from the four aforementioned images. To keep the data set balanced, the final EEG data on each specific participant was composed of 40 trials in which each object appears appeared 10 times.

For each subject, there were 40 trials, each lasting 5 s. Hence, each participant contributed a recording of 200 s. Since the sampling rate was 128 Hz, each subject contributed $25,600 = 128 \times 200$ sampling points, which meant the data set had $204,800$ sampling points in total.

10.1.3 *Online System*

In this section, we summarize our experience in developing a working prototype of Brain2Object. Figure 10.3 shows the working prototype in action. The graphic user interface is provided in Figure 10.4. The top of the interface shows the port number and baud rate of the IP printer. The IP address of the server which stores the pretrained model and makes the object recognition decision is also shown. The main body of the interface

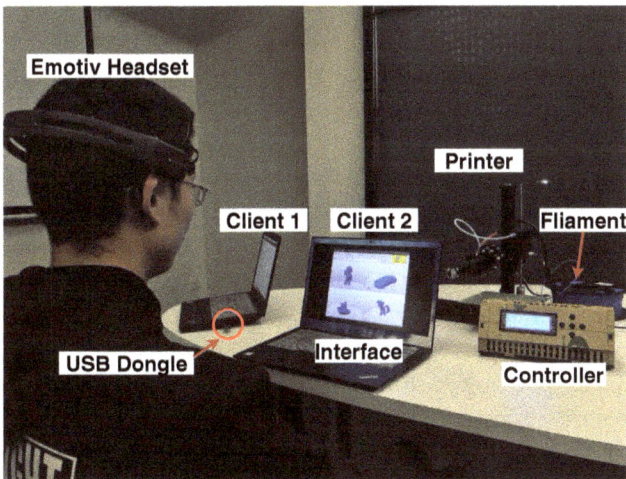

Fig. 10.3: Online testing scenario. The user's EEG signals are collected by Emotiv headset for recognition. The corresponding object will be printed by the 3D printer.

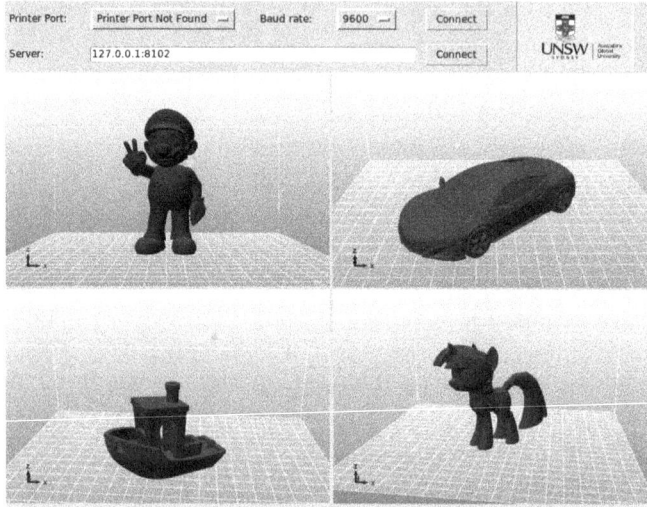

Fig. 10.4: User interface of the designed Brain2Object system. In the top frame, the interface shows the printer port, the baud rate, and the server IP address (with connection port). The main body of the interface displays four object models including Mario, the car, the boat, and Pinkie Pie pony.

displays object models for the four objects in our experiments, namely, Mario, the car, the boat, and Pinkie Pie Pony.

Figure 10.5 illustrates the operational workflow of the Brain2Object demonstrator. While the user is focusing on a target object (e.g., Pinkie Pie), the corresponding brain signals are collected by a properly mounted Emotiv headset and transmitted to client 1 over Bluetooth. Client 1 sends the EEG signal to the server over a TCP connection. The server loads the pretrained EEG recognition model and classifies the EEG signal as one of the four categories. The classification result is forwarded to the interface through client 2. The interface will highlight the selected object by changing the color of the other three objects to gray (the selected object remains blue). Simultaneously, the selected object is dispatched to the printer driver, which generates the corresponding G-code recognizable to the mechanical 3D printer. Finally, the G-code is sent to the 3D printer, which brings the object to life.

We used a Tronxy X1 desktop 3D printer with the following specifications: printer size: $220 \times 220 \times 250$ mm, build area: $150 \times 150 \times 150$ mm, MK10 extruder diameter: 1.75 mm, nozzle diameter: 0.4 mm, engraving accuracy: 0.1 mm, filament material: 1.75 mm polylactic acid (PLA).

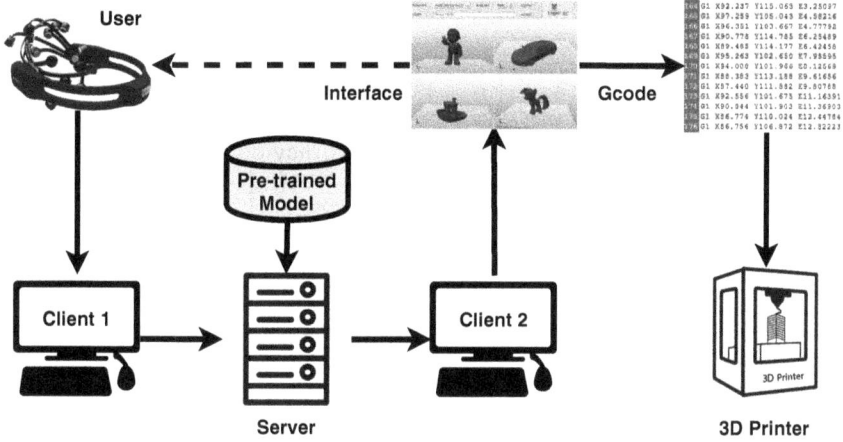

Fig. 10.5: Online workflow of Brain2Object. The user's EEG signals are collected and sent to the server through client 1. The server loads the pre-trained model to recognize the target object and spread to both the interface for showing the user feedback and the 3D printer for printing. The solid line denotes signal transmission while the slash line denotes feedback.

The physical 3D model can be transmitted from a computer to the printer or directly stored in minor SD card mounted on the printer.

The sampling frequency of the Emotiv headset is 128 Hz, which indicates that it can collect 128 sampling points per second. The pretrained recognition model requires that each sample have 14 sampling points and each sampling point correspond to a classification output. To achieve steadiness and reliability, the server will maintain a window of the last 10 classification outputs and a count of how many times each of the four objects has been recognized. The server will send the target to client 2 only if a specific target appears more than six times in this window. In this situation, the classification accuracy is higher than 90% although the latency is increased to about 2 s, which includes data collection time (1.1 s), recognition time (0.47 s), transmission time, and so on.

10.1.4 *Discussions*

First, the object repository in this section is limited. An ideal instantiation of Brain2Object should recognize any object the user observes. However, in this section, the object repository contained only four items. The scalability of the repository is constrained by the scalability of the learning algorithm, that is, the ability of multiclass classification algorithms to discriminate between a large number of classes. Classification accuracy falls dramatically reduces with an increased number of categories. In our pre-experiment, which for brevity is not presented in this section, in the offline test, the proposed approach achieved around 90% accuracy on binary classification using the Emotiv headset; however, the accuracy dropped to near 80% with three objects and to about 70% with four objects. One of the promising future scopes is to improve multiclass classification performance.

Additionally, the ideal printing system could automatically detect an object the subject was "thinking" of (without visual stimulation) in addition to any "observed" (with visual stimulation). However, the SNR of EEG signals (without visual stimulation) is much lower than those with visual stimulation. To enhance signal quality and to help the subject to concentrate on the target object, we adopted visual stimuli in our experiments; therefore, in the online demonstration phase, the corresponding object images were displayed on the monitor for the participants.

Furthermore, we observed that online performance was lower than offline analysis, which could be attributed to a number of reasons. First, the user's mental state and emotional fluctuations may affect the quality of the EEG signals. For instance, if the pretrained model was tuned based on the EEG data collected while the user was relaxed, the classification performance may be affected if the user is excited during the online phase. Second, the conductivity of the electrodes in the headset is not exactly invariant during the offline and online stage, which will impact the data quality. Third, subtle variations in the way the EEG headset is positioned on the subject's head (e.g., the placement of each of the electrodes) may also influence online decision making. Fourth, subjects often have difficulty maintaining concentration during signal acquisition.

10.2 Geometrical Shape Reconstruction

In this section, we demonstrate that EEG signals are sensitive to visual geometric shapes. Then, we present a novel and interesting application for reconstructing the geometric shape shown in the user's mind through accurate EEG decoding.

Since the advent of neuroscience, numerous studies have tried to recover perceived visual stimuli based on informative human brain activities (Seeliger *et al.*, 2018; Zhang *et al.*, 2019c). The development of decoding

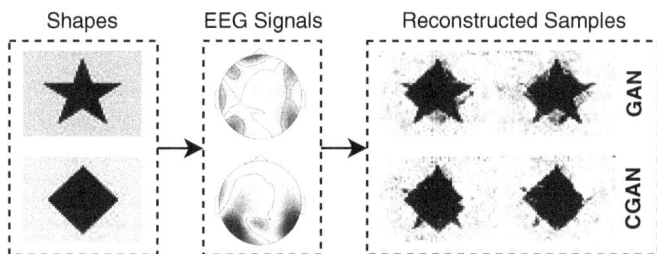

Fig. 10.6: Generated samples based on EEG signals evoked by geometric shapes. We see that the samples synthesized by traditional methods (e.g., GAN and CGAN) are blurry and lacking in realistic detail.

technologies of chaotic brain signals is thought to reveal functional mechanisms of brain neurons and may implement some fantastic ambitions such as mindreading (Zhang *et al.*, 2018h). Most of the existing work focuses on the use of fMRI to monitor brain activities by detecting changes associated with blood flow in various brain areas. However, fMRI-based image reconstruction faces several major challenges (Nishimoto *et al.*, 2011; Seeliger *et al.*, 2018). The temporal resolution of fMRI is low, constrained by the blood flow speed; the acquisition of fMRI requires a scanner which is prohibitively expensive; and the scanner is heavy, with poor portability (Zhang *et al.*, 2019c).

Thus, EEG has recently drawn much attention for its high temporal resolution, low cost, and high portability. Researchers have tried to exploit EEG signals to reconstruct visual stimuli (Kavasidis *et al.*, 2017; Palazzo *et al.*, 2017) through GANs. The current EEG-based synthesis methods can roughly present the visual stimuli but fail to capture the necessary detail. For example, as shown in Figure 10.6, clear geometric shapes are presented to the individual and the shapes are reconstructed based on the collected EEG data. The results demonstrate that the geometric shapes generated by traditional GAN and CGAN are blurry and lacking in realistic detail.

To address the aforementioned issues, we have conducted experiments to measure the individual's EEG oscillation evoked by various geometric shapes; in this section, we propose a novel framework in order to accurately decode the EEG signals and precisely synthesize the geometric shapes.

10.2.1 *EEG Signal Acquisition*

We conducted a local experiment with eight healthy participants, five males and three females aged 25 ± 3 years, which was approved by the UNSW ethics abroad (HC190315). Throughout the experiments, participants were required to sit in a comfortable armchair in front of a computer monitor. Five representative and widely seen geometric shapes (circle, star, triangle, rhombus, and rectangle) were presented to each subject. The experiment consisted of two sessions; each session had five trials. In each trial, the five geometric shapes were presented in random order, and each shape was displayed for 5 s, with 5 s of relaxation between each successive shape. The relaxation time between trials and sessions were 10 and 30 s, respectively. The EEG signals were collected through a portable Emotiv EPOC+ headset with 14 electrodes, and the sampling frequency was 128 Hz. Each EEG segment contained 10 continuous instances with 50% overlap. The data set was randomly divided into a training set (80%) and testing set (20%).

10.2.2　*Methodology*

Next, we first decoded the noninvasive EEG signals into an implicit representation (Section 10.2.2.1) and then proposed a modified GAN framework to generate the real shape which evoked the EEG signals (Section 10.2.2.2).

10.2.2.1　*EEG Feature Learning*

In the EEG feature learning process, we adopted a CNN structure to capture the latent distinguishable features from the collected EEG signals. Prior research has demonstrated that CNN is capable of learning informative features from noisy EEG data (Acharya *et al.*, 2018a; Zhang and Wu, 2019). Assuming the EEG sample pairs can be denoted by $\boldsymbol{E} = \{(\boldsymbol{E}_h, \boldsymbol{y}_h), h = 0, 1, \ldots, H\}$ where $\boldsymbol{E}_h \in \mathbb{R}^{M \times N}$ and $\boldsymbol{y}_h \in \mathbb{R}^5$ represent the EEG observations and the corresponding one-hot label, we focus on the decoding of five EEG patterns which correspond to five imagined object; thus, the number of labels is five. H denotes the number of EEG segments and M, N denotes the temporal and spatial resolution of each segment.

　　Figure 10.7 shows the workflow discriminative representation learning. The second to last fully connected layer, which is regarded as the learned representation, denoted by $\bar{\boldsymbol{E}}$, has d nodes and contains enough information to reconstruct the visual shape.

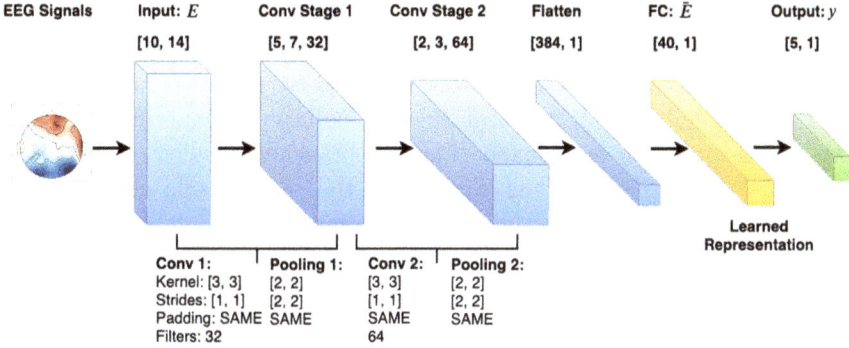

Fig. 10.7: Demonstration of discriminative EEG representation learning. The second to last layer with discriminative information $\bar{E}$ is selected as the learned representation. Each Conv stage contains a convolutional layer followed by a pooling layer. Only the basic hyperparameters are presented.

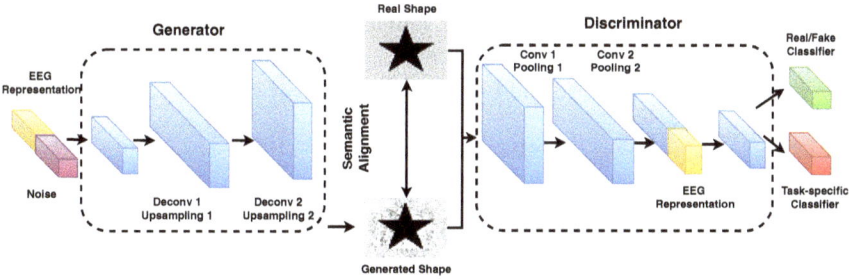

Fig. 10.8: Workflow for the proposed visual stimuli reconstruction framework. We adopted a semantic classifier in addition to the real/fake classifier in order to exploit the semantic information contained within the EEG samples. Moreover, a semantic regularization constraint is utilized to minimize the difference between the generated visual stimuli and the real one.

10.2.2.2 *Multitask Generation Model*

Overview. As shown in Figure 10.8, the proposed geometric shape generative framework contains two components: a generator and a discriminator.

The generator receives the learned discriminative EEG representation $\bar{E} \in \mathbb{R}^d$ along with a random sampled Gaussian noise $z \in \mathbb{R}^{d'}$ and produces a generated shape. We use EEG representations to improve quality of generated shapes while the Gaussian noise is adopted to preserve diversity. On the other hand, the discriminator receives the real shape and the generated fake shape. Inspired by ACGAN (Odena *et al.*, 2017), we designed a multitask discriminator containing two branches in which the first branch, like in a standard GAN, aimed at the recognition of the fake shapes and the second branch, an auxiliary task-specific classifier, attempted to classify each shape. The first branch was called the real/fake classifier, whereas the second was called the task-specific classifier. By adding the task-specific classifier, the discriminator was intended not only to be able to distinguish whether the shape was real or not but also to be able to categorize it. Consequently, the discriminator drove the distribution of the synthesized shapes not only approximate to the general distribution of the overall real shapes but also approximate to the distribution of a specific category. In addition, the learned EEG representations are also fed into the discriminator, as proposed in Mirza and Osindero (2014).

Architecture. The generator received the input vector which concatenated $\bar{\boldsymbol{E}}$ and $\boldsymbol{z}$, represented by $\boldsymbol{h_0} = \{\bar{\boldsymbol{E}} : \boldsymbol{z}\} \in \mathbb{R}^{d+d'}$, and attempted to map it to a meaningful shape. The generator was composed of one fully connected and two deconvolutional layers, each followed by an unsampling layer. $\boldsymbol{h_0}$ was first fed into the fully connected layer with $64(M+N)$ nodes:

$$\boldsymbol{h_1} = \sigma(\boldsymbol{w}\boldsymbol{h_0} + \boldsymbol{b}) \tag{10.14}$$

where $\boldsymbol{w}, \boldsymbol{b}$ and σ denote the weight, bias vector, and the sigmoid function, respectively. Then $\boldsymbol{h_1}$ was reshaped into $(M, N, 64)$, where 64 denotes the depth. To this end, $\boldsymbol{h_1}$ has a similar form but deeper depth, with the raw EEG segment $\boldsymbol{E_h}$, which is supposed to contain enough information to reconstruct the user's imagined shape. Afterward, $\boldsymbol{h_1}$ was sent to the first deconvolutional layer with 32 filters, kernel size $(5, 5)$, stride $(2, 2)$, and "SAME" padding method. The upsampling operation was the inverse operation of pooling and shared the same parameters. The second deconvolutional layer had one filter and a followed upsampling layer. We chose Tanh as the activation function, since it transforms the signals into the range $[-1, 1]$, which is the same range the real shape falls into. The synthesized shape $\boldsymbol{F}$ had shape $(4M, 4N)$. In accordance with our empirical experience, we set the shape size at four times the EEG raw segment in both width and height in order to obtain a better generation quality. The real geometric shape $\boldsymbol{R}$ was in grayscale with format $[4M, 4N]$. All the pixels were normalized into the range $[0, 1]$ by max–min normalization and then transformed to $[-1, 1]$ by:

$$\bar{\boldsymbol{R}} = 2\boldsymbol{R} - 1 \tag{10.15}$$

As shown in Figure 10.8, both $\bar{R}$ and F were fed into the discriminator, which had almost the same structure and hyperparameters as the discriminative representation learning model (Section 10.2.2.1). The input shape was flattened to a vector and then concatenated with the learned representation $\bar{E}$. The fully connected layer had 100 nodes. This discriminator design had two branches corresponding to two output layers. The output layer of the real/fake classifier had only one node representing the fake probability. As for the task-specific classifier, the output layer had five nodes corresponding to five different categories of geometrical shapes.

Loss Function. We present the loss functions in the proposed framework. For the generator, since we added a task-specific classifier, the loss function contained two components: one forces the discriminator to recognize the shape generated, whereas another urges the discriminator to recognize which category the shape belonged to. Thus, the log-likelihood loss function for the generator could be defined as (Odena *et al.*, 2017):

$$\mathcal{L}_g = E[\log P(C = y | X = F)] + E[\log(1 - D(G(y, \bar{E}, z)))] \qquad (10.16)$$

in which

$$F = G(y, \bar{E}, z) \qquad (10.17)$$

describes the generator G, and

$$P(S|X), P(C|X) = D(X) \qquad (10.18)$$

describes the real/fake classifier and task-specific classifier of the discriminator D, respectively. As for the discriminator, the loss function also contained two components coming separately from the two classifiers. The discriminator was supposed to filter out which shape was generated as well as to assign the shape to the correct class. The log-likelihood loss function $\mathcal{L}_d$ for the discriminator is:

$$
\begin{aligned}
\mathcal{L}_d = {}& E[\log P(S = \bar{R} | X = \bar{R})] + E[\log P(S = F | X = F)] \\
& + E[\log P(C = y | X = \bar{R})]
\end{aligned}
\qquad (10.19)
$$

In the above formula, y represents the class label; C, S denotes the predicted class and source, which are the classification results of the multitask generator; X denotes the shape fed into the discriminator; $P(S|X)$ denotes the probability distribution over the source S; and $P(C|X)$ denotes the probability distribution over the class label y.

10.2.2.3 *Semantic Alignment*

To this end, the geometric shape reconstruction model able to generate a batch of samples that have sufficient diversity but still lacked discriminability. In order to increase discriminability of the generated samples and make the samples more realistic, we propose a semantic alignment method. In particular, we added an additional constraint on the generator loss function aimed at reducing the distance between the real and the generated geometric shapes in order to make the synthesized shape more realistic and sharper.

The semantic distance can be measured by S_r:

$$S_r = \frac{1}{\sqrt{\bar{N}}} \sqrt{\sum_{i=0}^{\bar{N}} \sum_{j=0}^{\bar{N}} (\bar{R}_{i,j} - F_{i,j})^2} \tag{10.20}$$

where $\bar{N}$ denotes the number of pixels in the geometric sample and $\bar{N} = 4M \times 4N$. $\bar{R}_{i,j}$ and $F_{i,j}$ denote the pixels in the real and generated samples. In order to improve the performance of the generator, S_r is considered as a regularization of the generator loss. Thus, we update Equation (10.16) as:

$$\mathcal{L}_g = E[\log P(C = \boldsymbol{y}|X = \boldsymbol{F})] + E[\log(1 - D(G(\boldsymbol{y}, \bar{\boldsymbol{E}}, \boldsymbol{z})))] + \lambda S_r \tag{10.21}$$

where λ is a constant coefficient weighting the semantic regularization. If the alignment constraint too strong, the generated shapes may have less diversity. In this section, we set $\lambda = 0.01$ to make a tradeoff between the diversity and discriminability of the generated samples.

During the training, both $\mathcal{L}_g$ and $\mathcal{L}_d$ were optimized by the Adam optimizer. The learning rate was set as 0.0002 with an exponential decay rate of 0.5. In each epoch, the $\mathcal{L}_g$ and $\mathcal{L}_d$ were separately trained in turn. The proposed framework converged after 120 epochs and trended to overfitting after 160 epochs, thus, we adopted the early stopping strategy of breaking off the iteration at the 150th epoch.

10.2.3 *Evaluations*

10.2.3.1 *Qualitative Comparison*

We compared the quality of the generated shapes by the proposed method against the state-of-the-art models. As shown in Figure 10.9, we chose the most widely used generative models, including GAN, CGAN, and ACGAN as our baseline.

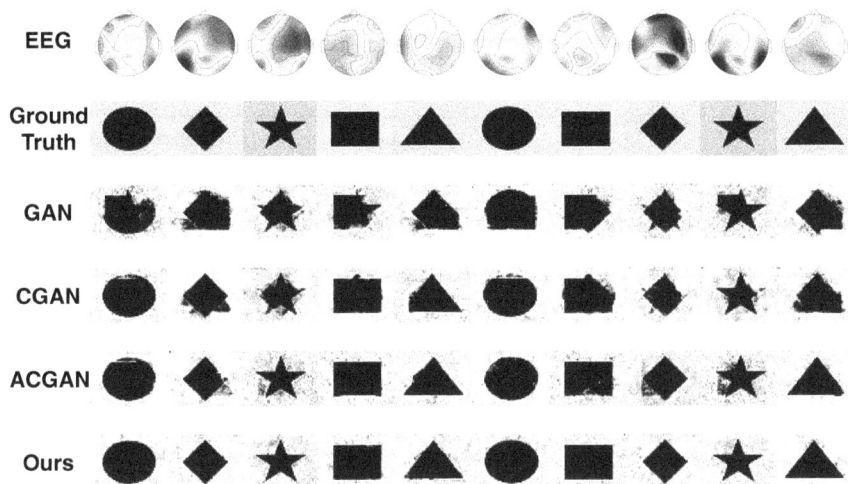

Fig. 10.9: Demonstration of the qualitative comparison. Our model can correctly reconstruct all the shapes that have the highest similarity to the ground truth.

GAN achieved promising results in many areas, especially in the shape field (Goodfellow *et al.*, 2014). On top of the basic GAN, CGAN (Mirza and Osindero, 2014) was proposed in order to add conditional information as a constraint, which was adopted in Kavasidis *et al.* (2017). Furthermore, ACGAN attempted to deeply exploit the informative sample labels to enhance the discriminability of D (Odena *et al.*, 2017). Our work, compared to ACGAN, proposed a semantic alignment method to constrain the distance between the synthesized shapes and the visual geometric shapes in order to better reflect reality.

As Figure 10.9 makes clear, our approach achieved the best shape quality. To be specific, the samples generated by GAN lacked clear edges, which is a typical mode collapse problem; meanwhile, it isn't hard to discern that most of the synthesized shapes contain mixed features. While CGAN performed better than GAN, as its shapes had a higher integrity, they also showed combined features, such as the star, which shows elements of the rhombus. ACGAN had the best results among the baseline models; it was able to learn most of the shapes' features and correctly reconstruct the shapes with only minor acceptable flaws. Our model, however, reconstructed all the shapes correctly and had the highest similarity to the ground truth.

Table 10.1: The quantitative comparison of inception score and inception accuracy.

Models	GAN	C-GAN	ACGAN	Ours
Inception Score	1.931	1.986	2.061	**2.178**
Inception Accuracy	0.43	0.67	0.79	**0.83**

10.2.3.2 *Quantitative Comparison*

We adopted the inception score and inception accuracy (Kavasidis *et al.*, 2017) for quantitative comparison. We built an inception network using the generated shapes as input in order to calculate the inception score, which measures the realism of the generated shapes. Furthermore, our work was intended to convert the specific EEG signals into the corresponding geometric shapes belonging to the specific labels. Thus, we adopted the performance of the task-specific classifier when the input data is F as inception accuracy in order to measure how precisely our model could generate shapes.

We conducted the quantitative analyses for the baselines and our proposed model. The results are presented in Table 10.1, which makes plain that our model achieved the highest inception score and inception accuracy, of 2.178 and 0.83, respectively.

10.2.4 *Discussions*

One major future challenge is the recovery of *unseen* geometric shapes; that is, how to decode the EEG signals evoked by a star, for instance, when the reconstruction model has never been trained on a star. One possible solution is to train a common generative model on a large class of basic geometric shapes (e.g., circle, ellipse, straight line, triangle, rectangle, and rhombus) in order to learn the latent features of each different shape and then approximate the unseen shape (e.g., star) based on the learned features.

Second, as a preliminary study, we only focused on the simple geometric shapes in this section. One future scope would thus be to consider more complicated geometric shapes. Another potential research direction would be to increase the number of geometric categories, since this section only evaluated five basic classes.

Chapter 11

Language Interpretation

Creating a valuable connection between the human brain and the outside world (Nguyen *et al.*, 2015), BCI systems allow people to communicate or interact with external devices such as wheelchairs or service robots, through their brain signals. Among the different types of brain signals, MI-EEG is especially popular and has demonstrated great potential in discerning different brain activities.

One of the most promising and widely discussed applications of EEG-based BCI systems enables people to type via direct brain control (Akram *et al.*, 2015). In this chapter, we present our attempt at building a brain-controlled typing system by enhancing the decoding accuracy of EEG signals via deep learning methods. We envision a real-world implementation of such a system able to interpret the user's thoughts to infer typing commands in real time. Those with motor disabilities would benefit greatly from such a system, as it would allow them to express their thoughts and interact with others.

We propose a novel hybrid deep neural network that combines the benefits of both CNN (Sak *et al.*, 2014) and RNN (Mikolov *et al.*, 2010) for effective EEG signal decoding. The network is capable of modeling high-level, robust, and salient feature representations hidden in the raw EEG signal streams and capturing complex relationships within data via stacking multiple layers of information-processing modules within a hierarchical architecture. We present an operational prototype of a brain-controlled typing system based on our proposed model that demonstrates the efficacy and practicality of this system. A video demonstration the system is also available.[1]

11.1   Methodology

11.1.1   *Overview*

Both CNN and RNN have been proved to be effective at EEG data decoding. Intuitively, we attempted to combine their advantages; however, our experiments demonstrated that the simple concatenation of temporal and spatial features cannot outperform either used alone. Therefore, we have designed a feature adaptation method that maps the stacked features to a new space and which can fuse the distinctive information obtained from the temporal and spatial features. Figure 11.1 illustrates the steps involved. The essential goal of our approach is to design a deep learning model that precisely classifies the user's successive intents based on EEG data. In summary, we propose a hybrid approach which contains three components: deep feature learning, feature adaption, and intent recognition.

11.1.2   *Deep Feature Learning*

We aimed to learn the representations of the user's typing intent signals. First we denoted the single input EEG signal as $E_{\bar{i}} \in \mathbb{R}^K$ ($K = 64$) where K is the number of dimensions in the raw EEG signal. Next, we fed $E_{\bar{i}}$ to the RNN structure and the CNN structure separately for temporal and spatial feature learning in parallel. Lastly, the learned temporal features X_t and the spatial features X_s were combined in the stacked feature X' for the feature adaption (Section 11.1.3).

11.1.2.1   *RNN Feature Learning*

We design an RNN model consisting of three components: one input layer, five hidden layers, and one output layer. The number of hidden layers is optimized by Orthogonal Array Tuning method. Through the experiential experiments, the hidden layers are designed to including three fully connected neural networks and two layers of Long Short-Term Memory (LSTM) (Zaremba *et al.*, 2014) (shown as rectangles in Figure 11.1) cells among the hidden layers. Assume a batch of input EEG data contains n_{bs}

---

[1]http://www.xiangzhang.info. Accessed Dec. 23, 2020.

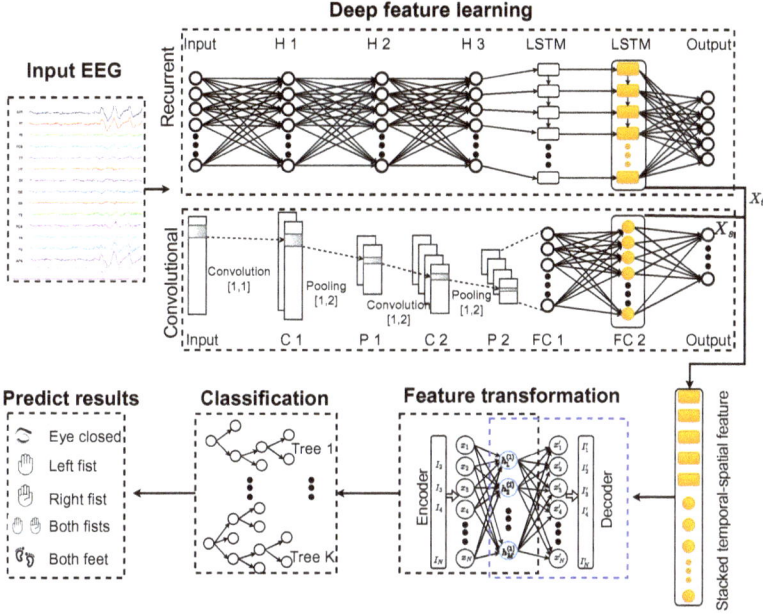

Fig. 11.1: The flowchart for the proposed language interpretation approach. The input raw EEG data is a single sample vector denoted by $E_{\bar{i}} \in \mathbb{R}^K$ (take $K = 64$ as an example). The C1 layer denotes the first convolutional layer, the C2 layer denotes the second convolutional layer, and so on. After the same fashion, the P1 layer denotes the first pooling layer, the FC1 layer denotes the first fully connected layer, the H1 layer denotes the first hidden layer. The stacked spatiotemporal feature is generated by the FC2 layer in the CNN and the H5 layer in the RNN.

(generally called batch size) EEG samples and the total input data has the 3-D shape as $[n_{bs}, 1, 64]$. Let the data in the ith layer ($i = 1, 2, \ldots, 7$) be denoted by $X_i^r = \{X_{ijk}^r | j = 1, 2, \ldots, n_{bs}, k = 1, 2, \ldots, K_i\}, X_i^r \in \mathbb{R}^{[n_{bs}, 1, K_i]}$, where j denotes the jth EEG sample and K_i denotes the number of dimensions in the ith layer.

Assume that the weights between layer i and layer $i+1$ can be denoted by $W_{i(i+1)}^r \in \mathbb{R}^{[K_i, K_{i+1}]}$, for instance, W_{12}^r describes the weight between layer 1 and layer 2. $b_i^r \in \mathbb{R}^{K_i}$ denotes the biases of ith layer. The data flow from the ith layer to the $i + 1$-th layer as follows:

$$X_{i+1}^r = sigmoid(X_i^r * W_{i,i+1}^r + b_i^r) \tag{11.1}$$

Please note the sizes of X_i^r, $W_{i,i+1}^r$ and b_i^r must match. For example, in Figure 11.1, the transformation between H1 layer and H2 layer, the sizes of $X_3^r, X_2^r, W_{[2,3]}$, and b_2^r are $[1, 1, 64], [1, 1, 64], [64, 64]$, and $[1, 64]$, respectively.

The fifth and sixth layers in the designed structure are LSTM layers, so the calculation in these layers are the same as Equation (3.8) to Equation (3.13). At last, we obtain the RNN prediction results X_7^r and employ cross-entropy as the cost function. The cost is optimized by the Adam optimizer algorithm (Kingma and Ba, 2014). X_6^r is the data in the second to last layer, which has a directly linear relationship with the output layer and the prediction results. If the prediction results have a high level of accuracy, X_6^r will be able to directly map to the sample label space and will produce the better representative of the input EEG sample. Therefore, we regard X_6^r as the temporal feature extracted by the RNN structure and call it X_t.

11.1.2.2 *CNN Feature Learning*

While RNN is good for exploring temporal (inter-sample) relevance, it is incapable of appropriately decoding spatial feature (intrasample) representations. We have designed a CNN structure to exploit the spatial connections between different features in each specific EEG signal. As shown in Figure 11.1, the proposed CNN was as follows: the input layer, the first convolutional layer, the first pooling layer, the second convolutional layer, the second pooling layer, the first fully connected layer, the second fully connected layer, and the output layer.

The input is the same EEG data as was fed into the RNN. The input EEG single sample $E_{\bar{i}}$ has the shape $[1, 64]$. Suppose the data in the ith layer $(i = 1, 2, \ldots, 8)$ is denoted by $X_i^c, X_i^c \in \mathbb{R}^{[1, K_i^c, d_i]}$, where K_i^c and d_i separately denote the dimension number and the depth in the ith layer. The data in the first layer has a depth of only 1 and $X_1^c = E$. We choose the convolutional filter with size $[1, 1]$ and the stride size $[1, 1]$ for the first convolution. The stride denotes the movement of the filter in x and y increments per slide. The padding method is selected as 'SAME' zero-padding, which results in the sample shape remaining constant in the convolution calculation. The depth of the EEG sample transfers to 2 through the first convolutional layer, so the shape of X_2^c is $[1, 64, 2]$. We have included a ReLU activation function designed to work on the convolutional results in our model.

The pooling layer is a nonlinear down-sampling transformation layer. There are several pooling options, with max pooling being the most popular (Nagi *et al.*, 2011). The max pooling layer scans through the inputs along with a sliding window with a designed stride. Then it outputs the maximum value in every subregion in which the window is scanned. The pooling layer reduces the spatial size of the input EEG features and also prevents overfitting. In the first pooling layer (the third layer of the CNN), we choose the $[1, 2]$ window and $[1, 2]$ stride. The maximum in each $[1, 2]$ window will be output to the next layer. The pooling does not change the depth, and the shape of X_3^c is $[1, 32, 2]$. Similarly, the second convolutional layer chooses $[1, 2]$ filter and $[1, 1]$ stride and results in a shape of $[1, 32, 4]$. The results are made nonlinear by the ReLU unit. The second pooling layer selects $[1, 2]$ window and $[1, 2]$ stride and obtains a shape of $[1, 16, 4]$.

In the fully connected layer, the high-level reasoning features, extracted by the prior convolutional and pooling layers, are unfolded to a flattened vector. For example, the data of the second pooling layer (X_5^c with shape $[1, 16, 4]$) is flattened to the vector with shape $[1, 64]$ (X_6^c). Then the output data can be calculated by following the regular neural network operation:

$$X_7^c = sigmoid(T(X_6^c)) \tag{11.2}$$

$$X_8^c = softmax(T(X_8^c)) \tag{11.3}$$

At last, we have the CNN results X_8^c with shape $[1, 5]$ and employ the cross-entropy as the cost function. The error is optimized by the Adam optimizer algorithm. X_7^c has a directly linear relationship with the output layer and the predicted results. Therefore, we regard X_7^c as the spatial feature extracted by the CNN structure and call it X_s. The proposed approach can automatically learn the distinguishable features from 1D EEG signals through the CNN structure. The order of the channels does not matter, as the training data set and the testing data set have the same order. No effort is needed to transfer the 1D data to 2D for spatial feature learning.

In summary, the temporal features X_t and the spatial features X_s are learned through the parallel RNN and CNN structures. Both of them have a direct linear relationship with the EEG sample label, which means that they represent the temporal and spatial features of the input EEG sample if both RNN and CNN have a high classification accuracy. Next, we combine the two feature vectors into a flattened stacked vector, $X' = \{X_t : X_s\}$.

11.1.3   Feature Adaptation

For better latent spatiotemporal representation learning, we have designed a feature adaptation method for mapping the stacked features to a correlative new feature space which can fuse the temporal and spatial features together and highlight their useful information.

To do so, we introduced the autoencoder layer (Nguyen *et al.*, 2015) to further interpret EEG signals, which is an unsupervised approach to learning effective features. The autoencoder is trained to learn a compressed and distributed representation for the stacked EEG feature X'. The input of autoencoder is the stacked temporal and spatial feature X'. Assume that h, $\acute{X}'$ denote the hidden layer and output layer data, respectively.

The data transformation procedure is described as follows:

$$\acute{X}' = sigmoid(T(sigmoid(T(X')))) \qquad (11.4)$$

The cost function measures the difference between X' and $\acute{X}'$ as the mean squared error (MSE), which is backpropagated to the algorithm to adjust the weights and biases. The error is optimized by the RMSPropOptimizer (Hinton *et al.*, 2012). The data in the hidden layer h is the transferred feature, which is output to the classifier. Finally, the XGBoost is employed (Chen and Guestrin, 2016) to classify the EEG streams. It fuses a set of classification and regression trees (i.e., CART) and extracts detailed information from the input data. It builds multiple trees, each with its leaves and corresponding scores.

11.2 Brain-Controlled Typing System

Based on the high classification accuracy for the proposed deep learning approach, in this section, we develop an online brain-controlled typing system (Figure 11.2) to convert users' thoughts to texts.[2] Compared with the state-of-the-art application (Speier *et al.*, 2017), the proposed method achieves a trade-off between several characteristics: noninvasive (low-cost, low-risk, and portable), completely functional (ability to input, cancel, delete, and confirm), high-speed typing, and full-dictionary.[3] The typing commands in the brain-controlled typing system are shown in Table 11.1.

---

[2]For brevity, the parameter settings, comparison with state-of-the-art models, and latency analysis are omitted. Please refer to Zhang *et al.* (2018h) for more details.

[3]The lack-dictionary functions such that after the user types in a character, for instance, it only has 6 choices ("A,E,H,I,O,R") for the next character instead of the full 26 choices. The alternative situation is the full-dictionary.

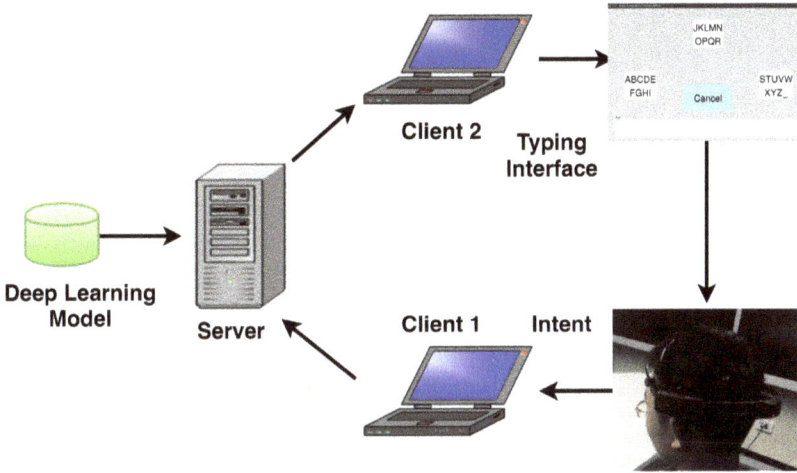

Fig. 11.2: Overview of the brain-controlled typing system. The user's typing intent is collected by the headset and sent to the server through client 1. The server uses the pretrained deep learning model to recognize the intent, which is used to control the typing interface through client 2. The server and clients are connected over TCP connections.

Table 11.1: The motor imagery tasks and labels and the corresponding typing commands in the brain-controlled typing system.

Data Set	Item	Task 1	Task 2	Task 3	Task 4	Task 5
EEGMMIDB	intent	eyes closed	left hand	right hand	both hands	both feet
	label	0	1	2	3	4
Emotiv	intent	up arrow	down arrow	left arrow	right arrow	eyes closed
	label	0	1	2	3	4
	Command	up	down	left	right	confirmation

The proposed system contains five components: EEG headset, client 1 (data collector), the server, client 2 (typing command receiver), and the typing interface. The user wears the Emotiv EPOC+ headset which collects EEG signals and sends the data to client 1 through a Bluetooth connection. The raw EEG signals are transported to the server over a TCP connection. The server feeds the incoming EEG signals to the pretrained deep learning model. The model produces a classification decision and converts it to the corresponding typing command, which is sent to client 2 over a TCP connection. The typing interface receives the command and manifests the appropriate typing action.

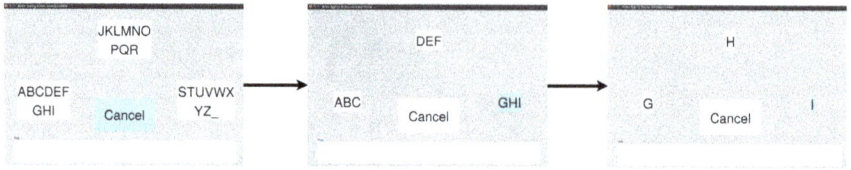

Fig. 11.3: The brain-controlled typing procedure for typing the character "I". First, select the left character block (containing the characters "ABCDEFGHI") in the *initial interface* and then confirm the selection to enter the corresponding *subinterface*; then select the right character block (containing only "GHI") in the *sub-interface* and confirm to jump to the *bottom interface*; last, select the right character block (containing only "I") and the character "I" appears in the display block after confirmation.

Specifically, the typing interface (Figure 11.3) can be divided into three levels: the initial interface, the subinterface, and the bottom interface. The interfaces all have a similar structure: three *character blocks* (distributed in the left, up, and down directions), a *display block*, and a *cancel button*. The display block shows the typed output and the cancel button is used to cancel the last operation. The typing system in total includes $27 = 3*9$ characters (the 26 letters of the English alphabet and the space bar) grouped into three character blocks (each of which contains nine characters) in the initial interface. In detail, there are three possible selections, and each of which leads to a specific subinterface which containing nine characters. Again, the $9 = 3*3$ characters are divided into three character blocks, each of which is connected to a bottom interface. In the bottom interface, each block represents only one character. For example, Figure 11.3 shows the procedure for typing the character "I".

In the brain-controlled typing system, there are five commands for controlling the interface: "left," "up," "right," "cancel," and "confirm." Each command corresponds to a specific motor imagery EEG category (Table 11.1). To type each single character, the interface must receive six commands. Consider typing the letter "I" for example (Figure 11.3). The sequence of commands to be entered is as follows: "left," "confirm," "right," "confirm," "right," "confirm." In our practical deployment, the sampling rate of Emotiv EPOC+ headset is set at 128 Hz, which means the server can receive 128 EEG recordings each second. Since the brainwave signal varies rapidly and is easily to be affected by noise, the EEG data stream is sent to the server *each half second*, which means that the server receives 64 EEG samples each time. The 64 EEG samples are classified by the deep learning framework and generate 64 categories of intents. We calculate the mode of

64 intents and regard the mode as the final intent decision. Furthermore, to achieve steadiness and reliability, the server sends a command to client 2 only if *three consecutive decisions remain consistent*. After the command is sent, the command list will be reset and the system will wait until the next three consistent decisions are made. Therefore, client 2 must wait for at least 1.5 s for each command and the entire process of typing each character necessarily takes at least 9 (6∗1.5) s. In other words, theoretically, the proposed brain-controlled typing system can achieve a maximum typing speed of $6.67 = 60/9$ characters per minute.

11.3 Discussion

The accuracy in the online mode is lower than that in an offline setting (over 95%), which may be attributed to a number of causes. First, the user's mental state and emotional fluctuations may affect the quality of the EEG signals. For example, a scenario in which the offline data set used to train the deep learning model was collected when the user was in an excited emotional state but was then applied in an online setting when the user was upset, would predictably result in low accuracy. Subtle variations in the way the EEG headset is positioned on the subject's head may also impact online decision making. Specifically, the position of each of the electrodes (e.g., the 14 electrodes in the Emotiv headset) on the scalp may vary between training and testing. Furthermore, EEG signals vary from person to person, which makes it difficult to construct a common model equally effective for all individuals. Part of our future work must be to identify the intraclass variabilities common to the activities of different subjects. Last but not least, several limitations are introduced by the intrinsic attributes of the headset. For instance, the headset used in our case study is too tight for the user to wear for longer than 30 min, and the conductivity of the wet electrodes decreases after prolonged use.

Chapter 12

Intent Recognition in Assisted Living

The EEG signals reflecting the activities of certain brain areas not requiring any initiative actions is an effective method for connecting the individuals with the outer world. In this chapter, we show how EEG-based BCI works in smart living scenarios. EEG-based intent control has shown promising performance in various applied fields. For instance, if a person suffering from ALS had only very limited physical capacities, she might be unable to communicate with the outer world, or incapable of performing most daily activities (e.g., turning on/off the light). It can be very difficult for her to use the usual platforms to control smart home appliances. EEG signals can be captured to recognize one's intent, as input used to communicate or interact with external smart devices such as wheelchairs or service robots in real-time BCI systems (Alomari *et al.*, 2014a).

The traditional EEG-based intent recognition approaches have limitations such as feature engineering being time consuming and having low accuracy. To address these challenges, we adopted a deep RNN model for intent recognition due to its excellent high-level distinctive feature learning ability to learn from raw data.

12.1 System Overview

As shown in Figure 12.1, the system consists of two components: the online component and the offline component. In the online component, raw EEG data, collected from subjects, are used to train a deep LSTM model. The model works directly on raw EEG data without any preprocessing, smoothing, filtering or feature extraction. The parameters in the deep

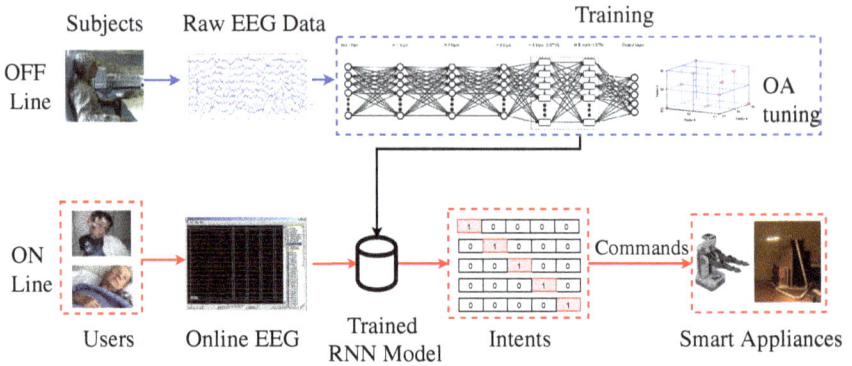

Fig. 12.1: Workflow of the BCI system in assisted living. OA tuning refers to orthogonal array tuning method.

learning model are optimized by the orthogonal array tuning method (Section 12.2). In the offline component, the user's intent (EEG signal) is sent to abovementioned pretrained RNN model for recognition. Learned intents may then be subsequently used to command devices, such as turning lights on/off or prompting a robot to serve a cup of water.

12.2   Orthogonal Array Tuning Method

12.2.1   *Overview*

Next, we very briefly introduce a hyperparameter tuning method, the OATM, which is very efficient in the context of deep learning. For further details, refer to Zhang *et al.* (2019a).

Deep learning faces a significant challenge in that the performance of the algorithm is highly dependent upon the selection of hyperparameters. Compared with traditional machine learning algorithms, deep learning more urgently requires hyperparameter tuning because deep neural networks: (1) have more hyperparameters requiring tuning; (2) have a higher dependency on the configuration of hyperparameters. Zhang *et al.* (2017b) report a dramatic fluctuation in the deep learning classification accuracy (from 32.2% to 92.6%) due solely to the selection of hyperparameters. Therefore, an effective and efficient hyperparameter tuning method is necessary.

However, most of the existing hyperparameter tuning methods involve some drawbacks. In particular, grid search traverses all the possible combinations of hyperparameters, which is a time-consuming and ad hoc process (Bergstra *et al.*, 2011). Random search, which was developed based on grid search, sets up a grid of hyperparameter values and selects random combinations to train the algorithm (Bergstra *et al.*, 2011); this method overcomes some disadvantages of grid search such as time consumption but introduces its own major disadvantage in that it cannot converge to the global optimum (Andradóttir, 2015). Randomly selected hyperparameter combinations cannot guarantee a reliable and competitive result. Apart from the manual tuning methods, automated tuning methods have grown more popular in recent years (Snoek *et al.*, 2012). Bayesian Optimization, the most widely used automated hyperparameter tunning, attempts to find the global optimum in a minimum number of steps. Nevertheless, the results of this method are sensitive to the parameters of the surrogate model and its performance is highly dependent upon the quality of the learning model (Calandra *et al.*, 2014).

Thus, we have adopted the OATM that achieves tradeoff between tuning time and competitive performance. To the best of our knowledge, ours is the first batch of work to adopt adopting orthogonal array for parameter tuning in deep learning.

12.2.2 *OATM Workflow*

OATM is a systematic and statistical method, and its principle is to compare the dependent variable resulting from a combination of independent variables. It selects only certain representative combinations rather than all combinations for testing. In this method, the independent variable is called the "factor," and different values of the factor are called "levels." For instance, if the program has three factors, each of which has three levels, which are represented by a cube with 27 nodes (each node represents one combination of hyperparameters), OATM chooses only nine representative groups of parameters to optimize the selection. As shown in Figure 12.2, A_1, A_2, A_3 represent three levels of factor A, factors B, C likewise represent three each (each factor is assumed to be statistically independent of the others). The nine circled nodes are the nine groups selected by OATM. Each edge (in total 27 edges) in the cube has one circled node and each face (in total nine faces) has three circled nodes.

The OATM procedure is as follows:

- **Step 1:** Build the factor-level (F-L) table. Determine the number of factors to be tuned and the number of levels for each factor. The

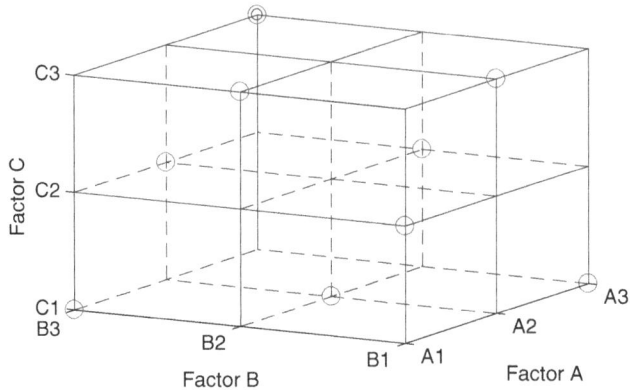

Fig. 12.2: OATM selection.

levels should be determined by experience and the literature. We further assume each factor has the same number of levels.[1]

- **Step 2:** Construct the Orthogonal Array Tuning table. The constructed table should obey the usual basic composition principles. (See the link below for some commonly used tables.[2]) The Orthogonal Array Tuning table is marked as $L_M(h^k)$ which has k factors, h levels, and, in total, M rows.

- **Step 3:** Run the experiments with the hyperparameters obtained by consulting the Orthogonal Array Tuning table.

- **Step 4:** Range analysis. This is the key step. Based on the experimental results from the previous step, employ the range analysis method to analyze the results and determine the optimal levels and importance of each factor. The importance of a factor is determined by its influence on the experimental results. Note that the range analysis optimizes each factor and combines the optimal levels, which means that the optimized hyperparameter combination is not restricted to the values given in the existing Orthogonal Array table.

- **Step 5:** Run the experiment with the optimal hyperparameters and report the final results.

---

[1] For the sake of simplicity, we assume that all of the factors have the same number of levels. Advanced knowledge covering more complex situations can be found in Taguchi (1987).

[2] https://www.york.ac.uk/depts/maths/tables/taguchi_table.htm. Accessed Aug. 19, 2017.

Table 12.1: Intents and corresponding labels and functions in case studies.

Intent	Label	Robot (Case 1)	Household Appliance (Case 2)
Eyes Closed	1	Walk-forward	Turn on blue LEDs
Left Fist	2	Turn left	Turn on white LED
Right Fist	3	Turn right	Turn on yellow LED
Both Fists	4	Grasp	Turn on red LED
Both Feet	5	Release	Turn on all LEDs

12.3 Deployment

Next, we demonstrate the efficiency of the EEG-based intent recognition system through two assisted living applications. The intents and commands used in two scenarios are shown in Table 12.1.

12.3.1 *Mind-Controlled Mobile Robot*

In this case, an assistive robot is used to move a canned beverage from the kitchen table to table in the living room. The simulation platform is the Gazebo toolbox[3] and the robot's controlling program is powered by Robot Operating System (ROS).[4] The PR2 mobile robot was choosen as the assistive robot for its smooth disk base and flexible hand design. The simulation environment is depicted in Figure 12.3. The video demo can be found in the link below.[5]

The input EEG signals are sampled from the EEGMMIDB data set. The path is shown in Figure 12.3, which is designed to cause the EEG data to drive PR2 to complete its service task. Starting from the vicinity of the kitchen table, the PR2 robot walks forward and holds out its hand to grasp the canned beverage. Then it turns back, walks along the path to the table in living room, and opens its hand to set the can down on the table. This sequence demonstrates that the robot can precisely grasp and release target objects in accordance with a plan conceived in the subject's mind. The robot executes five actions in response to five commands as described in Table 12.1.

---

[3]http://gazebosim.org/. Accessed Feb. 14, 2020.
[4]http://www.ros.org/. Accessed Mar. 23, 2020.
[5]https://www.youtube.com/watch?v=VZYX1095Vkc. Accessed Dec. 14, 2020.

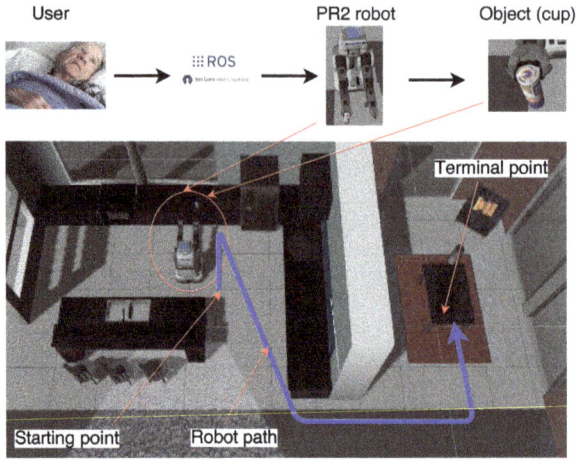

Fig. 12.3: Mind-controlled PR2 assistive robot performs a daily task: getting a glass of water from the kitchen and setting it on a table in living room.

12.3.2   *Mind-Controlled Appliances*

The most common scenario in an assisted living environment would be controlling household appliances. In this case, we controlled four LEDs ON/OFF through intent. LED commands corresponding to specific intents are listed in Table 12.1. For each command, the corresponding LED lights up for 2 s and then turns off. Such a test is conducted 10 times for a total of 80 commands, which our model accomplished with *100%* accuracy, which indicating that EEG-based mind control has the potential to become significant in smart households in the future.

Chapter 13

Patient-Independent Neurological Disorder Detection

13.1 Introduction

In this chapter, we present a method of detecting neurological disorders (e.g., epilepsy) through deep learning algorithms. Epilepsy is a chronic neurological disorder that affects about 1% of the world's population (Detti *et al.*, 2018; Hosseini *et al.*, 2017a). Such abnormal brain activity may cause seizures, unusual behavior and sensations, and sometimes loss of awareness. An accurate and timely diagnosis for epileptic seizure is crucial to its financial toll as well as its toll on quality of life. Spontaneous EEG is the most common method for diagnosing seizures (Zhang *et al.*, 2019c) but faces several challenges.

Challenges. First, most of the existing epileptic seizure detection methods focus on patient-dependent scenario (Altaf *et al.*, 2015; Detti *et al.*, 2018), rarely consider the patient-independent alternative.[1] The former refers to detection of a patient's epileptic seizure by learning from his own historical records, and the latter to learning from the records of other patients. Patient-dependent methods can achieve high accuracy because the training and testing samples are gathered from the same source and have a similar distribution. Patient-dependent algorithms have been widely investigated over the past decade due to their good performance for recorded patients;

---

[1]In some of the literature, the patient-dependent/independent methods are called patient-specific/nonspecific.

however, their performance is deficient for new patients since the data distributions from unknown patients is necessarily much more diverse. By contrast, patient-independent methods present an advance in alerting potential patients but are easily corrupted by interpatient noise (e.g., gender, age, and epileptic type). The majority of existing studies fail to eliminate such noise, and, to the best of our knowledge, few studies try to eliminate the interpatient corruption by modeling the common features from the training samples. Thus, in this chapter, we provide a robust method capable of learning the common seizure patterns while also mitigating the influence of interpatient factors.

Second, the noninvasive EEG signals have a low SNR and are susceptible to both subjective factors (e.g., emotion and fatigue) and environmental factors (e.g., noise) (Goh *et al.*, 2018; Li *et al.*, 2015a). To discover a latent and informative representation of the signals, most traditional seizure detection methods perform feature engineering on raw EEG signals, which is time-consuming and highly reliant upon expertise (Boashash and Ouelha, 2016). Recently, deep learning has shown great success for many research topics (e.g., computer vision, natural language processing, and brain–computer interface) because of its excellent automatic feature learning ability (Thodoroff *et al.*, 2016). Therefore, deep learning algorithms have great potential for reducing the negative effects of manual feature engineering by automatically capturing the key differences between epileptic and nonepileptic seizure patterns.

Motivation. The aim of this chapter is to develop an automatic representation learning algorithm that is dependent upon neurological status (e.g., seizure or normal) yet independent of personal status (e.g., age and gender). The representation is learned from noninvasive EEG signals and will be used to diagnose epileptic seizure. As mentioned above, a main challenge in patient-independent seizure feature learning is that the learning process can be easily affected by patient-specific details (e.g., age, gender, and emotional state). In order to eliminate such influences, we propose an adversarial method (Zhang *et al.*, 2019e) for decomposing the EEG signals into a seizure-related component and a patient-related component.

In short, the intuitive idea of this chapter is to separate the input raw EEG signals (denoted by $\mathcal{E}$) into two parts: a seizure-related component, denoted by $\mathcal{S}$, that contains informative descriptions of the seizure state and is insensitive to the patient identity; and a patient-related component, denoted by $\mathcal{P}$, that contains only patient identity information and is insensitive to the seizure state. The following two constraints should be satisfied during this process. First, the sum of seizure-related component $\mathcal{S}$ and patient-related component $\mathcal{P}$ should be equal to the original EEG signal $\mathcal{E}$. Second, the decomposed $\mathcal{S}$ and $\mathcal{P}$ should contain pure and informative seizure and patient features, respectively.

Contributions. We propose an adversarial representation learning framework to construct a robust patient-independent detection algorithm that eliminates the corruption caused by interpatient noise. The framework harnesses both a deep generative model and a convolutional discriminative model to capture informative seizure and patient representations.

Most of the existing seizure diagnosis methods (whether traditional or deep learning models) focus on the patient-dependent scenario, which means that the EEG training and testing samples are from the same patient or group of patients. Although very limited studies mention patient-independent validation as has Orosco *et al.* (2016), they tend to extract the general representations through enhancing the training data set instead of designing a robust patient-independent feature learning algorithm. As a result, the diagnosis model can handle intrapatient factors but it fails to eliminate interpatient noise. Clinical treatment requires the inclusion of situations of higher complexity (patient-independent situations) in which the testing patient is unseen in the training stage. It is much more difficult to perform accurate seizure diagnosis considering personal factors such as chronological age, gender, characteristics, and state of health. In summary, proposing a robust method to deal with the patient-independent challenge is necessary.

In this chapter, we propose a robust framework targeting patient-independent epileptic seizure detection. The proposed approach efficiently captures the seizure-specific representations directly from the raw EEG signals.

13.2 Methodology

13.2.1 *Overview*

The proposed patient-independent seizure representation learning method comprises three sections (Figure 13.1): EEG decomposition, seizure diagnosis, and patient detection. Joint adversarial training is used concurrently on all three. To satisfy the first constraint, the EEG signals are mapped to

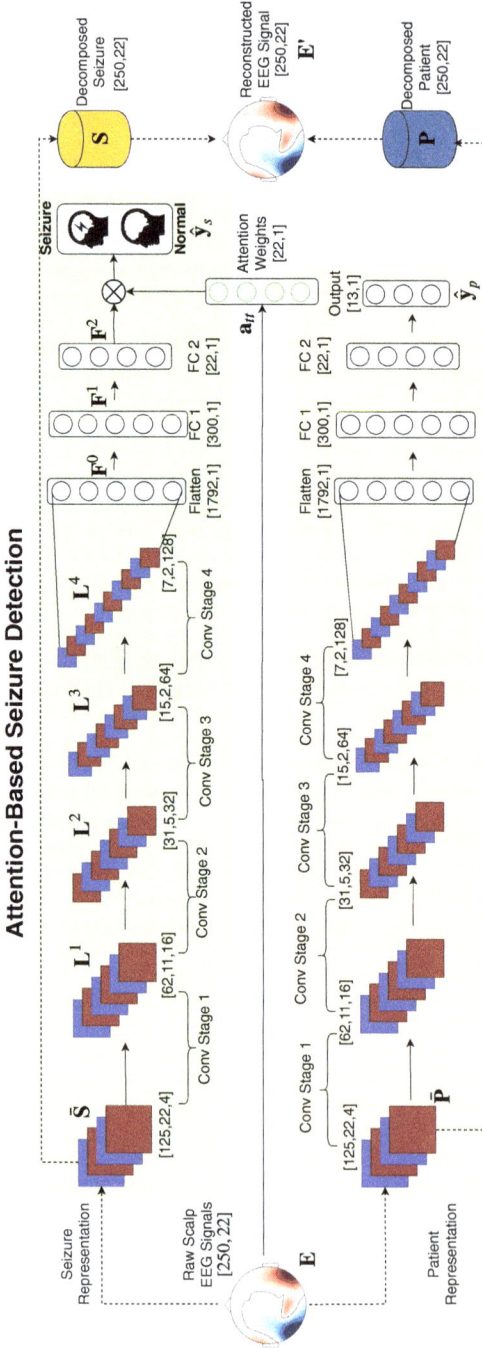

Fig. 13.1: Workflow of the patient-independent epileptic seizure diagnosis method. The raw EEG signals are first embedded into a latent space and decomposed into seizure and patient representations. Ultimately, the implicate representations will be used to reconstruct the seizure and patient components, respectively. In the meantime, the representations are sent through the seizure diagnosis and patient detection classifiers for diagnose of the user's seizure stage and recognition of the user's identity. The key branch is the seizure diagnosis, while the patient detection branch is adopted to separate the patient-related information from the input EEG signals. The dashed lines represent the data flow of EEG decomposition.

a latent space and decomposed into the seizure and patient representation, corresponding to the seizure state and personal information, respectively. Next, the two latent representations are used to reconstruct the original EEG signal. If the reconstructed signal is equal to the input signal, the learned seizure and patient representations are regarded as effective features of the decomposed EEG data.

In addition, to satisfy the second constraint that guarantees the purification of the decomposed representations, we designed a seizure-based diagnosis section and a patient classifier section to follow the generation of latent representations. The seizure diagnosis section, which follows the seizure representation, aims to distinguish the epileptic seizure state from a normal state. Seizure diagnosis is set up to repeat learning seizure representation until the diagnosis achieves a competitive performance. Similarly, the patient classifier is designed to loop until it reliably recognizes the user's identity, while a well-performed diagnosis ensures the purity of the patient representation and vice versa.

The EEG decomposition section focuses more on the reconstruction performance of the decomposed components than the components' purity; conversely, the seizure diagnosis section and patient classifier sections consider the purity of the two components rather than the effectiveness of the representations. As a whole, the two aspects are optimized in opposite directions, which taken together thus form an adversarial training situation. It should be noted that the patient classifier is only adopted to complete the adversarial model and improve the signal decomposition performance.

Furthermore, we introduce an attention mechanism with our seizure diagnosis section in order to pay more attention to the EEG electrodes. The effectiveness of the attention mechanism has been demonstrated by a number of research topics like the natural language process (Zhou *et al.*, 2016). The modified attention-based model is adopted in order to discover which electrodes make the greatest contributions to the seizure identifying data.

13.2.2 *EEG Decomposition*

Assume the gathered EEG signals are denoted by $\mathcal{E} = \{E_i \in \mathbb{R}^{M \times N}\}$, in which each EEG sample E_i has M time series and N channels. We suppose the EEG signals can be decomposed into the seizure-related component $\mathcal{S} = \{S_i \in \mathbb{R}^{M \times N}\}$ and the patient-related component $\mathcal{P} = \{P_i \in \mathbb{R}^{M \times N}\}$, described as[2]

$$\boldsymbol{E} = \boldsymbol{S} + \boldsymbol{P} \tag{13.1}$$

Nevertheless, $\mathcal{S}$ and $\mathcal{P}$ cannot be directly calculated using traditional methods. To surmount this challenge, we proposed a novel decomposition model based on deep neural networks. In particular, the $\boldsymbol{E}$ is transformed into a latent space by a convolutional operation followed by a max pooling layer:

$$\bar{\boldsymbol{S}} = ReLU(\boldsymbol{w_s} \circledast \boldsymbol{E} + \boldsymbol{b_s}) \tag{13.2}$$

$$\bar{\boldsymbol{P}} = ReLU(\boldsymbol{w_p} \circledast \boldsymbol{E} + \boldsymbol{b_p}) \tag{13.3}$$

where $\bar{\boldsymbol{S}}, \bar{\boldsymbol{P}} \in \mathbb{R}^{J \times K \times H}$ denote the latent seizure and patient representation, J and K denote the rows and columns of the learned representation, respectively, and H denotes the number of convolutional filters. The $\circledast$ denotes the convolution operation and $\boldsymbol{w_s}, \boldsymbol{b_s}, \boldsymbol{w_p}, \boldsymbol{b_p}$ represent the corresponding weights and biases. In each convolutional layer, the activation function is ReLU and the padding method is "SAME."

Then, in order to ensure the representations contained enough discriminative information, we attempted to reconstruct the EEG signals through deconvolutional operations:

$$\boldsymbol{S} = ReLU(\boldsymbol{w'_s} \circledast \bar{\boldsymbol{S}} + \boldsymbol{b'_s}) \tag{13.4}$$

$$\boldsymbol{P} = ReLU(\boldsymbol{w'_p} \circledast \bar{\boldsymbol{P}} + \boldsymbol{b'_p}) \tag{13.5}$$

where $\boldsymbol{S}$ and $\boldsymbol{P}$ represented the decomposed seizure- and patient-related components, which had the same shape as the input EEG signal $\boldsymbol{E}$. The reconstructed signal $\boldsymbol{E'} \in \mathbb{R}^{M \times N}$ was calculated by combining the two decomposed signals

$$\boldsymbol{E'} = w_1 \boldsymbol{S} + w_2 \boldsymbol{P}; w_1 + w_2 = 1 \tag{13.6}$$

where w_1 and w_2 are weights.

To guarantee the decomposition performance and separate the useful seizure-related information while eliminating the intrapatient corruption, we forced the reconstructed signal $\boldsymbol{E'}$ to approximate the original signal $\boldsymbol{E}$ by minimizing the mean square error (MSE, aka ℓ_2) loss between them.

---

[2]For simplicity, we omit the subscript.

The MSE distance has been widely applied to design the reconstruction loss function due to its simplicity and efficiency (Amaral *et al.*, 2013). The MSE loss indeed outperforms the cross-entropy and integral probability metrics — e.g., Wasserstein loss (Arjovsky *et al.*, 2017)) — in our preliminary experiments. The MSE loss function can be calculated as

$$\mathcal{L}_D = \|\boldsymbol{E} - \boldsymbol{E'}\|_2 \tag{13.7}$$

The learned latent seizure $\bar{\boldsymbol{S}}$ and patient representation $\bar{\boldsymbol{P}}$, compared to the decomposed components $\boldsymbol{S}$ and $\boldsymbol{P}$, have lower dimension and represent high-level features. Thus, the learned representations are used for the corresponding seizure diagnosis and patient detection in the next process.

13.2.3 *Attention-Based Seizure Diagnosis*

The weights of different EEG electrodes are variable in seizure diagnosis (Temko *et al.*, 2011); thus, we employed an attention mechanism to learn the channel importance and to pay differing amounts of attention to various signal channels. The attention mechanism, as mentioned above, allows modeling of dependencies among input sequences (Vaswani *et al.*, 2017) and has proved to be successful in some research topics (Zhou *et al.*, 2016). Meanwhile, the excellent latent feature learning ability of CNNs has been widely used in research areas such as computer vision (Zhang *et al.*, 2017a) and the natural language process (Zhang *et al.*, 2016b). Thus, we propose an attention-based CNN algorithm to automatically extract the distinctive information from the received seizure representation $\bar{\boldsymbol{S}}$, which has shape $[J, K, H]$.

As shown in Figure 13.1, the attention-based seizure diagnosis contains four convolutional stages: the flattening layer, two fully connected layers (FCs) and an output layer. Each convolutional stage contains a convolutional layer and a max pooling layer. The input layer receives the learned seizure feature $\bar{\boldsymbol{S}}$ and is sent to the convolutional layers:

$$\boldsymbol{C^i} = ReLU(\boldsymbol{w_s^i} \circledast \boldsymbol{L^{i-1}} + \boldsymbol{b_s^i}), \quad i \in \{1, 2, 3\} \tag{13.8}$$

where $\boldsymbol{C^i}$ and $\boldsymbol{L^i}$ denote the ith convolutional and pooling layer, respectively. If $i = 1$, $\boldsymbol{L^0}$ equals to the input $\bar{\boldsymbol{S}}$; otherwise, the pooling layer can be measured by:

$$\boldsymbol{L_u^i} = \max_{u \in \mathcal{U}}\{\boldsymbol{C_u^i}\}, \quad i \in \{1, 2, 3\} \tag{13.9}$$

where $\mathcal{U}$ represents the max pooling perception field and u denotes the uth element in $\mathcal{U}$.

The fourth pooling layer $\boldsymbol{L^4}$ is flattened to 1D vector $\boldsymbol{F^0}$ and then fed into the FC layers:

$$\boldsymbol{F^i} = \sigma(\boldsymbol{w_f^i}\boldsymbol{F^{i-1}} + \boldsymbol{b_f^i}), \quad f \in \{1, 2, 3\} \tag{13.10}$$

where σ denotes a nonlinear activation function (e.g., sigmoid) and $\boldsymbol{F^i}$ denotes the ith FC layer.

The predicted seizure state $\hat{y}_s$ is related to the last FC layer $\boldsymbol{F^2}$ and the learned attention weights $\boldsymbol{a_{tt}}$:

$$\hat{y}_s = \boldsymbol{F^2} \cdot \boldsymbol{a_{tt}} \tag{13.11}$$

where $\cdot$ denotes the dot production operation, whereas $\boldsymbol{a_{tt}}$ is directly learned from the EEG sample as follows:

$$\boldsymbol{a_{tt}} = sigmoid(\boldsymbol{w_a}\boldsymbol{E} + \boldsymbol{b_a}) \tag{13.12}$$

where $\boldsymbol{w_a}$ and $\boldsymbol{b_a}$ denote the corresponding parameters.

The cross-entropy loss function of seizure diagnosis $\mathcal{L}_s$ is defined as:

$$\mathcal{L}_s = -(\boldsymbol{y_s}log(\boldsymbol{p}(\hat{\boldsymbol{y}}_{\boldsymbol{s}})) + (1 - \boldsymbol{y_s})log(1 - \boldsymbol{p}(\hat{\boldsymbol{y}}_{\boldsymbol{s}}))) \tag{13.13}$$

where $\boldsymbol{y_s}$ and $\boldsymbol{p}(\hat{\boldsymbol{y}}_{\boldsymbol{s}})$ denote the ground truth of the seizure state and the predicted probability of the patient in seizure state, respectively.

13.2.4 *Patient Detection*

The architecture of the patient detection is almost identical to the attention-based seizure diagnosis except for two significant differences: (1) the patient detection component does not require an attention module; (2) the patient detection component performs multiclass rather than binary classification. Similar to $\mathcal{L}_s$, we can measure the multiclass cross-entropy loss function $\mathcal{L}_p$ of the patient identity detection classifier as:

$$\mathcal{L}_p = -\sum_{c=1}^{C}(\boldsymbol{y_p}log(\hat{\boldsymbol{y}}_{\boldsymbol{p}})) \tag{13.14}$$

where $\boldsymbol{y_p}$, $\hat{\boldsymbol{y}}_{\boldsymbol{p}}$, and C denote the ground truth of patient identity, the predicted identity, and the overall number of training patients, respectively.

13.2.5 Training Details

To this end, we have three loss functions which are the reconstruction loss $\mathcal{L}_D$ in signal decomposition, the classification loss $\mathcal{L}_s$, and $\mathcal{L}_p$ in the attention-based seizure diagnosis and patient detection, respectively.

We propose an adversarial training strategy to jointly train all the loss functions:

$$\mathcal{L} = \mathcal{L}_D + \mathcal{L}_s + \mathcal{L}_p + \ell_2 \qquad (13.15)$$

which assures all losses work on the gradient at the same time and converge to a trade-off position that balances the reconstruction performance and the purification of the decomposed signals. The ℓ_2 represents the ℓ_2 norm with a coefficient of 0.0001 to prevent overfitting. Then, since the seizure diagnosis is the most crucial component in this approach, we train the $\mathcal{L}_s$ once more to raise its priority. To sum up, in each training epoch, the $\mathcal{L}$ and $\mathcal{L}_s$ are optimized once in turn. We adopt the Adam optimizer with learning rate e^{-4} for both loss functions. The algorithm is trained for 250 epochs. And a dropout layer with a 0.8 keep rate is added to the flattening layer in both classifiers to prevent overfitting.

13.3 Discussions

In this chapter, we have proposed a novel deep learning–based neurological disorder detection system aimed at patient-independent epileptic seizure diagnosis that is demonstrated to have a competitive performance and high interpretability. Moreover, it also sheds light on subject-independent brain signal representation learning via signal decomposition.

The proposed system is also able to protect the patients' privacy. By extracting the seizure representation during the EEG decomposition phase, the learned $\bar{S}$ comes to contain person-related information, making it difficult to infer the patient's identity from the $\bar{S}$. In real-world deployment, the patients' privacy is thus protected.

Next, we'll discuss the preliminary challenges and potential directions for future work. The proposed adversarial framework achieved an average patient-independent diagnosis accuracy of around 80%, which, while outperforming the state-of-the-art baselines, is not reliable enough for clinical deployment. One feasible solution is to adopt ensemble strategies such as voting, bootstrap sampling, or aggregation. Taking voting strategy as an example, we could select multiple EEG segments and predict the seizure state independently. Then, the final decision would be obtained by the majority vote from all the prediction results. Such strategy may increase the latency, but it also improves the prediction accuracy. From the perspective of time efficiency, if we have five EEG segments where each segment costs 1 s, the data collection latency will be 5 s, which is acceptable in a real-world scenario. From the perspective of prediction performance, suppose the original probability of correct classification p is 0.8; then, after voting from all of the EEG segments, for example, five segments, the probability becomes $1 - [C_5^4 p(1-p)^4 - C_5^5(1-p)^5] = 0.9933$, which is satisfactory in most the application scenarios. Thus, we can choose an appropriate number of segments to assure the model achieves better detection accuracy while preserving efficiency in real-world cases.

Furthermore, the proposed method contains a large set of hyperparameters (such as the convolutional filter size), which may lead to a heavy tuning workload. One possible method is to adopt a hyperparameter optimization algorithm — for example, OATM (Zhang *et al.*, 2019a) — to learn the optimal hyperparameter settings efficiently.

Chapter 14

Future Directions and Conclusion

14.1 Future Directions

Although deep learning has improved the performance of BCI systems, technical and usability challenges remain. The technical challenges concern the classification ability in complex brain signal scenarios, whereas the usability challenges concern limitations to large-scale real-world deployment. In this section, we introduce these challenges and share some thoughts regarding potential solutions.

14.1.1 *General Framework*

In this book, we have introduced several types of brain signals (e.g., spontaneous EEG, ERP, fMRI) and the deep learning models that have been applied for each type. One promising research direction for deep learning–based brain signal research is to develop a generalized framework capable of handling various brain signals regardless of the number of channels used for signal collection, sample dimensions (e.g., 1D or 2D samples), or stimulus types (e.g., visual or audio), and so on. A generalized framework would require two key capabilities: the attention mechanism and the ability to capture latent features, the former to guarantee the framework could focus on the most valuable parts of input signals and the latter to enable the framework to capture the distinctive and informative features.

The attention mechanism could be implemented based on attention scores or utilize various machine learning algorithms, such as reinforcement learning. The attention scores could be inferred from the input data and work as weights to help direct the framework's attention. Reinforcement learning has been shown to be able to find the most valuable features through a policy search (Zhang *et al.*, 2018d). CNN is the most suitable structure for capturing features in various levels and ranges. In the future, CNN could be used as a fundamental feature learning tool, integrated with suitable attention mechanisms to form a general classification framework.

14.1.2 *Subject-Independent Classification*

Until now, most brain signal classification tasks have focused on subject-dependent scenarios, where the training and the testing sets come from the same individual. Another future direction is to focus on accomplishing subject-independent classification so that the testing data need never appear in the training set. High-performance subject-independent classification is necessary for the wide application of BCI systems in the real world.

One possible method forward achieving this goal is to build a personalized model with transfer learning. A personalized effective model can adopt a transductive parameter transfer approach to construct individual classifiers and to learn a regression function that maps the relationship between data distribution and classifier parameters (Zheng and Lu, 2016). Another potential solution is mining the subject-independent components from the input data. The input data can be decomposed into two parts: a subject-dependent component, which depends on the subject, and a subject-independent component, common to all subjects. A hybrid multi-task model can work on two tasks simultaneously, e.g., person identification and class recognition. A well-trained, converged model ought to extract the subject-independent features in a class recognition task.

14.1.3 *Semi-Supervised and Unsupervised Classification*

The performance of deep learning models is highly dependent upon the size of the training data set; expensive and time-consuming manual labeling is required to provide abundant class labels in a wide range of scenarios, such as sleep EEGs. While supervised learning requires both observations and labels for the training stage, unsupervised learning only need observations without labels and semi-supervised learning requires observations and partial labels (Jia *et al.*, 2014). Therefore, the latter two methods are more suitable for problems with little ground truth data available.

Zhang *et al.* have proposed an adversarial variational embedding (AVAE) framework that combines a VAE++ model (as a high-quality generative model) and a semi-supervised GAN (as a posterior distribution learner) (Zhang *et al.*, 2019e) for robust and effective semi-supervised learning, while Jia *et al.* (2014) have proposed a semi-supervised framework created by leveraging label information in feature extraction and integrating unlabeled information to regularize the supervised training.

Two methods may enhance unsupervised learning: one is to employ crowdsourcing to label the unlabeled observations; the other is to leverage unsupervised domain adaption learning to align the distribution of source signals and target signals with a linear transformation.

14.1.4 *Hardware Portability*

Limited hardware portability has prevented the widespread application of BCI systems in the real world. In most scenarios, users would prefer to use small, comfortable, even wearable brain signal hardware to collect brain signals and to control appliances and assistive robots.

Currently, there are three types of EEG collection equipment: non-portable equipment, the portable headset, and ear-EEG sensors. The nonportable equipment (e.g., Neuroscan, Biosemi) has a high sampling frequency, number of channels, and high signal quality but is expensive, making it suitable mainly for physical examinations conducted in hospitals. Portable headsets (e.g., Neurosky, Emotiv EPOC) have 1–14 channels and 128–256 sampling rate but may cause discomfort after extended use. The ear-EEG sensors, which are attached to the outer ear, have garnered increasing attention recently but remain confined mostly to the laboratory setting (Pacharra *et al.*, 2017). The ear-EEG platform comprises a set of electrodes placed inside each ear canal, together with additional electrodes in the concha of each ear (Mikkelsen *et al.*, 2015). cEEGrids, to the best of our knowledge, is the only commercial ear-EEG. It consists of multichannel sensor arrays placed around the ear using an adhesive and is even more expensive.[1] A promising future direction is to improve usability by developing a cheaper (i.e., under $200) and more comfortable (e.g., can be worn longer than 3 hr without discomfort) wireless ear-EEG device.

---

[1]https://uol.de/psychologie/abteilungen/ceegrid. Accessed Sep. 2, 2021.

14.2   Conclusion

Finally, we would like to share some insights into the field of deep learning–based BCI. This field has a bright future, both in academia and industry. Bridging the human mind and the outer world, brain signals could be widely useful in a range of real-world scenarios (e.g., health care, public security, and entertainment) for not only those with disabilities but also for healthy individuals. Indeed, deep learning techniques have already and will continue to play a central role in the brain signal decoding which is the core component of BCI systems. With this book, we have hopefully helped our readers to grasp the basics of BCI systems and deep learning techniques as well as the advanced deep learning algorithms and their real-world BCI applications.

Bibliography

Abdelfattah, S. M., Abdelrahman, G. M., and Wang, M. (2018). Augmenting the size of eeg datasets using generative adversarial networks, in *2018 International Joint Conference on Neural Networks (IJCNN)* (IEEE), pp. 1–6.

Abdulkader, S. N., Atia, A., and Mostafa, M.-S. M. (2015). Brain computer interfacing: Applications and challenges, *Egyptian Informatics Journal* **16**, 2, pp. 213–230.

Acar, E., Bingol, C. A., Bingol, H., Bro, R., and Yener, B. (2007). Seizure recognition on epilepsy feature tensor, in *EMBS* (IEEE), pp. 4273–4276.

Acharya, U. R., Oh, S. L., Hagiwara, Y., Tan, J. H., and Adeli, H. (2018a). Deep convolutional neural network for the automated detection and diagnosis of seizure using eeg signals, *Computers in Biology and Medicine* **100**, pp. 270–278.

Acharya, U. R., Oh, S. L., Hagiwara, Y., Tan, J. H., Adeli, H., and Subha, D. P. (2018b). Automated eeg-based screening of depression using deep convolutional neural network, *Computer Methods and Programs in Biomedicine* **161**, pp. 103–113.

Adeli, H., Ghosh-Dastidar, S., and Dadmehr, N. (2007). A wavelet-chaos methodology for analysis of eegs and eeg subbands to detect seizure and epilepsy, *IEEE Transactions on Biomedical Engineering* **54**, 2, pp. 205–211.

Akram, F., Han, S. M., and Kim, T.-S. (2015). An efficient word typing p300-bci system using a modified t9 interface and random forest classifier, *Computers in Biology and Medicine* **56**, pp. 30–36.

Al-kaysi, A. M., Al-Ani, A., and Boonstra, T. W. (2015). A multichannel deep belief network for the classification of eeg data, in *International Conference on Neural Information Processing* (Springer), pp. 38–45.

Al-Naffakh, N., Clarke, N., Li, F., and Haskell-Dowland, P. (2017). Unobtrusive gait recognition using smartwatches, in *2017 International Conference of the Biometrics Special Interest Group (BIOSIG)* (IEEE), pp. 1–5.

Alhagry, S., Fahmy, A. A., and El-Khoribi, R. A. (2017). Emotion recognition based on eeg using lstm recurrent neural network, *Emotion* **8**, 10.

Almogbel, M. A., Dang, A. H., and Kameyama, W. (2018). Eeg-signals based cognitive workload detection of vehicle driver using deep learning, in *Advanced Communication Technology (ICACT), 2018 20th International Conference on* (IEEE), pp. 256–259.

Alomari, M. H., AbuBaker, A., Turani, A., Baniyounes, A. M., and Manasreh, A. (2014a). Eeg mouse: A machine learning-based brain computer interface, *International Journal of Advanced Research in Computer Science and Applications* **5**, 4, pp. 193–198.

Alomari, M. H., Baniyounes, A. M., and Awada, E. A. (2014b). Eeg-based classification of imagined fists movements using machine learning and wavelet transform analysis, *International Journal of Advancements in Electronics and Electrical Engineering* **3**, 3, pp. 83–87.

Altaf, M. A. B., Zhang, C., and Yoo, J. (2015). A 16-channel patient-specific seizure onset and termination detection soc with impedance-adaptive transcranial electrical stimulator, *IEEE Journal of Solid-State Circuits* **50**, 11, pp. 2728–2740.

Amaral, T., Silva, L. M., Alexandre, L. A., Kandaswamy, C., Santos, J. M., and de Sá, J. M. (2013). Using different cost functions to train stacked auto-encoders, in *2013 12th Mexican International Conference on Artificial Intelligence* (IEEE), pp. 114–120.

Andradóttir, S. (2015). A review of random search methods, in *Handbook of Simulation Optimization* (Springer), pp. 277–292.

Ang, K. K., Chin, Z. Y., Wang, C., Guan, C., and Zhang, H. (2012). Filter bank common spatial pattern algorithm on bci competition iv datasets 2a and 2b, *Frontiers in Neuroscience* **6**, p. 39.

Ansari, A. H., Cherian, P. J., Caicedo, A., Naulaers, G., De Vos, M., and Van Huffel, S. (2018). Neonatal seizure detection using deep convolutional neural networks, *International Journal of Neural Systems*, p. 1850011.

Antoniades, A., Spyrou, L., Martin-Lopez, D., Valentin, A., Alarcon, G., Sanei, S., and Took, C. C. (2018). Deep neural architectures for mapping scalp to intracranial eeg, *International Journal of Neural Systems*, p. 1850009.

Antoniades, A., Spyrou, L., Took, C. C., and Sanei, S. (2016). Deep learning for epileptic intracranial eeg data, in *Machine Learning for Signal Processing (MLSP), 2016 IEEE 26th International Workshop on* (IEEE), pp. 1–6.

Anumanchipalli, G. K., Chartier, J., and Chang, E. F. (2019). Speech synthesis from neural decoding of spoken sentences, *Nature* **568**, 7753, p. 493.

Arjovsky, M., Chintala, S., and Bottou, L. (2017). Wasserstein generative adversarial networks, in *International Conference on Machine Learning (ICML)*, pp. 214–223.

Attia, M., Hettiarachchi, I., Hossny, M., and Nahavandi, S. (2018). A time domain classification of steady-state visual evoked potentials using deep recurrent-convolutional neural networks, in *Biomedical Imaging (ISBI 2018), 2018 IEEE 15th International Symposium on* (IEEE), pp. 766–769.

Atum, Y., Pacheco, M., Acevedo, R., Tabernig, C., and Manresa, J. B. (2019). A comparison of subject-dependent and subject-independent channel selection strategies for single-trial p300 brain computer interfaces, *Medical & Biological Engineering & Computing* **57**, 12, pp. 2705–2715.

Aungsakun, S., Phinyomark, A., Phukpattaranont, P., and Limsakul, C. (2011). Robust eye movement recognition using eog signal for human-computer interface, in *International Conference on Software Engineering and Computer Systems* (Springer), pp. 714–723.

Aznan, N. K. N., Bonner, S., Connolly, J. D., Moubayed, N. A., and Breckon, T. P. (2018). On the classification of ssvep-based dry-eeg signals via convolutional neural networks, *arXiv preprint arXiv:1805.04157*.

Ba, J., Mnih, V., and Kavukcuoglu, K. (2014). Multiple object recognition with visual attention, *arXiv preprint arXiv:1412.7755*.

Bahdanau, D., Chorowski, J., Serdyuk, D., Brakel, P., and Bengio, Y. (2016). End-to-end attention-based large vocabulary speech recognition, in *Acoustics, Speech and Signal Processing (ICASSP), 2016 IEEE International Conference on* (IEEE), pp. 4945–4949.

Baltatzis, V., Bintsi, K.-M., Apostolidis, G. K., and Hadjileontiadis, L. J. (2017). Bullying incidences identification within an immersive environment using hd eeg-based analysis: A swarm decomposition and deep learning approach, *Scientific Reports* **7**, 1, p. 17292.

Bandt, S. K., Roland, J. L., Pahwa, M., Hacker, C. D., Bundy, D. T., Breshears, J. D., Sharma, M., Shimony, J. S., and Leuthardt, E. C. (2017). The impact of high grade glial neoplasms on human cortical electrophysiology, *PloS One* **12**, 3, p. e0173448.

Bashar, M. K., Chiaki, I., and Yoshida, H. (2016). Human identification from brain eeg signals using advanced machine learning method eeg-based biometrics, in *Biomedical Engineering and Sciences (IECBES), 2016 IEEE EMBS Conference on* (IEEE), pp. 475–479.

Bashashati, A., Fatourechi, M., Ward, R. K., and Birch, G. E. (2007). A survey of signal processing algorithms in brain–computer interfaces based on electrical brain signals, *Journal of Neural Engineering* **4**, 2, p. R32.

Bashivan, P., Kar, K., and DiCarlo, J. J. (2019). Neural population control via deep image synthesis, *Science* **364**, 6439, p. eaav9436.

Bashivan, P., Rish, I., and Heisig, S. (2016a). Mental state recognition via wearable eeg, *arXiv preprint arXiv:1602.00985*.

Bashivan, P., Rish, I., Yeasin, M., and Codella, N. (2016b). Learning representations from eeg with deep recurrent-convolutional neural networks, *ICLR*.

Bashivan, P., Yeasin, M., and Bidelman, G. M. (2015). Single trial prediction of normal and excessive cognitive load through eeg feature fusion, in *Signal Processing in Medicine and Biology Symposium (SPMB), 2015 IEEE* (IEEE), pp. 1–5.

Behncke, J., Schirrmeister, R. T., Burgard, W., and Ball, T. (2018). The signature of robot action success in eeg signals of a human observer: Decoding and visualization using deep convolutional neural networks, in *Brain-Computer Interface (BCI), 2018 6th International Conference on* (IEEE), pp. 1–6.

Belitski, A., Farquhar, J., and Desain, P. (2011). P300 audio-visual speller, *Journal of Neural Engineering* **8**, 2, p. 025022.

Bergstra, J. S., Bardenet, R., Bengio, Y., and Kégl, B. (2011). Algorithms for hyper-parameter optimization, in *NeurIPS 24*, pp. 2546–2554.

Bigdely-Shamlo, N., Mullen, T., Kothe, C., Su, K.-M., and Robbins, K. A. (2015). The prep pipeline: Standardized preprocessing for large-scale eeg analysis, *Frontiers in Neuroinformatics* **9**, p. 16.

Biswal, S., Kulas, J., Sun, H., Goparaju, B., Westover, M. B., Bianchi, M. T., and Sun, J. (2017). Sleepnet: Automated sleep staging system via deep learning, *arXiv preprint arXiv:1707.08262*.

Boashash, B. and Ouelha, S. (2016). Automatic signal abnormality detection using time-frequency features and machine learning: A newborn eeg seizure case study, *Knowledge-Based Systems* **106**, pp. 38–50.

Bruzzone, L. and Marconcini, M. (2009). Domain adaptation problems: A dasvm classification technique and a circular validation strategy, *IEEE Transactions on Pattern Analysis and Machine Intelligence* **32**, 5, pp. 770–787.

Calandra, R., Gopalan, N., Seyfarth, A., Peters, J., and Deisenroth, M. P. (2014). Bayesian gait optimization for bipedal locomotion, in *Learning and Intelligent Optimization*, pp. 274–290.

Callisaya, M. L., Blizzard, L., Schmidt, M. D., McGinley, J. L., Lord, S. R., and Srikanth, V. K. (2009). A population-based study of sensorimotor factors affecting gait in older people, *Age and Ageing* **38**, 3, pp. 290–295.

Cao, J., Guo, Y., Wu, Q., Shen, C., and Tan, M. (2018). Adversarial learning with local coordinate coding, *The International Conference of Machine Learning (ICML)*.

Carabez, E., Sugi, M., Nambu, I., and Wada, Y. (2017). Identifying single trial event-related potentials in an earphone-based auditory brain-computer interface, *Applied Sciences* **7**, 11, p. 1197.

Cavanagh, P. *et al.* (1992). Attention-based motion perception, *Science* **257**, 5076, pp. 1563–1565.

Cecotti, H. (2017). Convolutional neural networks for event-related potential detection: Impact of the architecture, in *Engineering in Medicine and Biology Society (EMBC), 2017 39th Annual International Conference of the IEEE* (IEEE), pp. 2031–2034.

Cecotti, H., Eckstein, M. P., and Giesbrecht, B. (2014). Single-trial classification of event-related potentials in rapid serial visual presentation tasks using supervised spatial filtering, *IEEE Transactions on Neural Networks and Learning Systems* **25**, 11, pp. 2030–2042.

Cecotti, H. and Graser, A. (2011). Convolutional neural networks for p300 detection with application to brain-computer interfaces, *IEEE Transactions on Pattern Analysis and Machine Intelligence* **33**, 3, pp. 433–445.

Cecotti, H. and Ries, A. J. (2017). Best practice for single-trial detection of event-related potentials: Application to brain-computer interfaces, *International Journal of Psychophysiology* **111**, pp. 156–169.

Chai, R., Ling, S. H., San, P. P., Naik, G. R., Nguyen, T. N., Tran, Y., Craig, A., and Nguyen, H. T. (2017). Improving eeg-based driver fatigue classification using sparse-deep belief networks, *Frontiers in Neuroscience* **11**, p. 103.

Chai, X., Wang, Q., Zhao, Y., Liu, X., Bai, O., and Li, Y. (2016). Unsupervised domain adaptation techniques based on auto-encoder for non-stationary eeg-based emotion recognition, *Computers in Biology and Medicine* **79**, pp. 205–214.

Chambon, S., Galtier, M. N., Arnal, P. J., Wainrib, G., and Gramfort, A. (2018). A deep learning architecture for temporal sleep stage classification using multivariate and multimodal time series, *IEEE Transactions on Neural Systems and Rehabilitation Engineering*.

Chan, W., Jaitly, N., Le, Q. V., Vinyals, O., and Shazeer, N. M. (2017). Speech recognition with attention-based recurrent neural networks, US Patent 9,799,327.

Chen, K., Yao, L., Wang, X., Zhang, D., Gu, T., Yu, Z., and Yang, Z. (2018). Interpretable parallel recurrent neural networks with convolutional attentions for multi-modality activity modeling, *International Joint Conference on Neural Networks (IJCNN)*.

Chen, T. and Guestrin, C. (2016). Xgboost: A scalable tree boosting system, in *Proceedings of the 22nd ACM SIGKDD International Conference on Knowledge Discovery and Data Mining* (ACM), pp. 785–794.

Chiappa, K. H. (1997). *Evoked Potentials in Clinical Medicine* (Lippincott Williams & Wilkins).

Chiarelli, A. M., Croce, P., Merla, A., and Zappasodi, F. (2018). Deep learning for hybrid eeg-fnirs brain–computer interface: Application to motor imagery classification, *Journal of Neural Engineering* **15**, 3, p. 036028.

Chin, Z. Y., Ang, K. K., Wang, C., Guan, C., and Zhang, H. (2009). Multi-class filter bank common spatial pattern for four-class motor imagery bci, in *2009 Annual International Conference of the IEEE Engineering in Medicine and Biology Society* (IEEE), pp. 571–574.

Chorowski, J. K., Bahdanau, D., Serdyuk, D., Cho, K., and Bengio, Y. (2015). Attention-based models for speech recognition, in *NeurIPS*, pp. 577–585.

Chu, L., Qiu, R., Liu, H., Ling, Z., Zhang, T., and Wang, J. (2017). Individual recognition in schizophrenia using deep learning methods with random forest and voting classifiers: Insights from resting state eeg streams, *arXiv preprint arXiv:1707.03467*.

Chuang, J., Nguyen, H., Wang, C., and Johnson, B. (2013). I think, therefore i am: Usability and security of authentication using brainwaves, in *International Conference on Financial Cryptography and Data Security* (Springer), pp. 1–16.

Cichy, R. M., Khosla, A., Pantazis, D., and Oliva, A. (2017). Dynamics of scene representations in the human brain revealed by magnetoencephalography and deep neural networks, *Neuroimage* **153**, pp. 346–358.

Cichy, R. M., Khosla, A., Pantazis, D., Torralba, A., and Oliva, A. (2016). Comparison of deep neural networks to spatio-temporal cortical dynamics of human visual object recognition reveals hierarchical correspondence, *Scientific Reports* **6**, p. 27755.

Ciresan, D. C., Meier, U., Gambardella, L. M., and Schmidhuber, J. (2011). Convolutional neural network committees for handwritten character classification, in *Document Analysis and Recognition (ICDAR), 2011 International Conference on* (IEEE), pp. 1135–1139.

Cola, G., Avvenuti, M., Musso, F., and Vecchio, A. (2016). Gait-based authentication using a wrist-worn device, in *Proceedings of the 13th International Conference on Mobile and Ubiquitous Systems: Computing, Networking and Services* (ACM), pp. 208–217.

Dai, M., Zheng, D., Na, R., Wang, S., and Zhang, S. (2019). Eeg classification of motor imagery using a novel deep learning framework, *Sensors* **19**, 3, p. 551.

Dang-Vu, T. T., Schabus, M., Desseilles, M., Albouy, G., Boly, M., Darsaud, A., Gais, S., Rauchs, G., Sterpenich, V., Vandewalle, G. *et al.* (2008). Spontaneous neural activity during human slow wave sleep, *Proceedings of the National Academy of Sciences* **105**, 39, pp. 15160–15165.

Deng, L. (2014). A tutorial survey of architectures, algorithms, and applications for deep learning, *APSIPA Transactions on Signal and Information Processing* **3**.

Derawi, M. and Voitenko, I. (2014). Fusion of gait and ecg for biometric user authentication, in *Biometrics Special Interest Group (BIOSIG), 2014 International Conference of the* (IEEE), pp. 1–4.

Detti, P., de Lara, G. Z. M., Bruni, R., Pranzo, M., Sarnari, F., and Vatti, G. (2018). A patient-specific approach for short-term epileptic seizures prediction through the analysis of eeg synchronization, *IEEE Transactions on Biomedical Engineering* **66**, 6, pp. 1494–1504.

Dong, H., Supratak, A., Pan, W., Wu, C., Matthews, P. M., and Guo, Y. (2018). Mixed neural network approach for temporal sleep stage classification, *IEEE Transactions on Neural Systems and Rehabilitation Engineering* **26**, 2, pp. 324–333.

Du, L.-H., Liu, W., Zheng, W.-L., and Lu, B.-L. (2017). Detecting driving fatigue with multimodal deep learning, in *Neural Engineering (NER), 2017 8th International IEEE/EMBS Conference on* (IEEE), pp. 74–77.

Duan, L., Bao, M., Miao, J., Xu, Y., and Chen, J. (2016). Classification based on multilayer extreme learning machine for motor imagery task from eeg signals, *Procedia Computer Science* **88**, pp. 176–184.

Dubbelink, K. T. O., Felius, A., Verbunt, J. P., Van Dijk, B. W., Berendse, H. W., Stam, C. J., and Delemarre-van de Waal, H. A. (2008). Increased resting-state functional connectivity in obese adolescents; a magnetoencephalographic pilot study, *PLoS One* **3**, 7, p. e2827.

Elisha, A. E., Garg, L., Falzon, O., and Di Giovanni, G. (2017). Eeg feature extraction using common spatial pattern with spectral graph decomposition, in *Computing Networking and Informatics (ICCNI), 2017 International Conference on* (IEEE), pp. 1–8.

Emek-Savaş, D. D., Güntekin, B., Yener, G. G., and Başar, E. (2016). Decrease of delta oscillatory responses is associated with increased age in healthy elderly, *International Journal of Psychophysiology* **103**, pp. 103–109.

Fahimi, F., Zhang, Z., Goh, W. B., Lee, T.-S., Ang, K. K., and Guan, C. (2019). Inter-subject transfer learning with an end-to-end deep convolutional neural network for eeg-based bci, *Journal of Neural Engineering* **16**, 2, p. 026007.

Farwell, L. A. and Donchin, E. (1988). Talking off the top of your head: Toward a mental prosthesis utilizing event-related brain potentials, *Electroencephalography and Clinical Neurophysiology* **70**, 6, pp. 510–523.

Fernández-Varela, I., Athanasakis, D., Parsons, S., Hernández-Pereira, E., and Moret, V. (2018). Sleep staging with deep learning: A convolutional model, in *Proceedings of the European Symposium on Artificial Neural Networks, Computational Intelligence and Machine Learning (ESANN 2018)*.

Fraiwan, L. and Lweesy, K. (2017). Neonatal sleep state identification using deep learning autoencoders, in *Signal Processing & its Applications (CSPA), 2017 IEEE 13th International Colloquium on* (IEEE), pp. 228–231.

Fraschini, M., Hillebrand, A., Demuru, M., Didaci, L., and Marcialis, G. L. (2015). An eeg-based biometric system using eigenvector centrality in resting state brain networks, *IEEE Signal Processing Letters* **22**, 6, pp. 666–670.

Friedman, J. H. (2001). Greedy function approximation: A gradient boosting machine, *Annals of Statistics*, pp. 1189–1232.

Frydenlund, A. and Rudzicz, F. (2015). Emotional affect estimation using video and eeg data in deep neural networks, in *Canadian Conference on Artificial Intelligence* (Springer), pp. 273–280.

Gao, W., Guan, J.-a., Gao, J., and Zhou, D. (2015a). Multi-ganglion ann based feature learning with application to p300-bci signal classification, *Biomedical Signal Processing and Control* **18**, pp. 127–137.

Gao, Y., Lee, H. J., and Mehmood, R. M. (2015b). Deep learninig of eeg signals for emotion recognition, in *Multimedia & Expo Workshops (ICMEW), 2015 IEEE International Conference on* (IEEE), pp. 1–5.

Garg, P., Davenport, E., Murugesan, G., Wagner, B., Whitlow, C., Maldjian, J., and Montillo, A. (2017). Automatic 1d convolutional neural network-based detection of artifacts in meg acquired without electrooculography or electrocardiography, in *Pattern Recognition in Neuroimaging (PRNI), 2017 International Workshop on* (IEEE), pp. 1–4.

Gers, F. A. and Schmidhuber, E. (2001). Lstm recurrent networks learn simple context-free and context-sensitive languages, *IEEE Transactions on Neural Networks* **12**, 6, pp. 1333–1340.

Gers, F. A., Schmidhuber, J., and Cummins, F. (1999). Learning to forget: Continual prediction with lstm.

Ghasedi Dizaji, K., Wang, X., and Huang, H. (2018). Semi-supervised generative adversarial network for gene expression inference, in *The 24th ACM SIGKDD International Conference on Knowledge Discovery & Data Mining* (ACM).

Givens, G. H., Beveridge, J. R., Lui, Y. M., Bolme, D. S., Draper, B. A., and Phillips, P. J. (2013). Biometric face recognition: From classical statistics to future challenges, *Wiley Interdisciplinary Reviews: Computational Statistics* **5**, 4, pp. 288–308.

Glauner, P. O. (2015). Comparison of training methods for deep neural networks, *arXiv preprint arXiv:1504.06825*.

Goh, S. K., Abbass, H. A., Tan, K. C., Al-Mamun, A., Thakor, N., Bezerianos, A., and Li, J. (2018). Spatio–spectral representation learning for electroencephalographic gait-pattern classification, *IEEE Transactions on Neural Systems and Rehabilitation Engineering* **26**, 9, pp. 1858–1867.

Goldberger, A. L., Amaral, L. A., Glass, L., Hausdorff, J. M., Ivanov, P. C., Mark, R. G., Mietus, J. E., Moody, G. B., Peng, C.-K., and Stanley, H. E. (2000). Physiobank, physiotoolkit, and physionet: Components of a new research resource for complex physiologic signals, *Circulation* **101**, 23, pp. e215–e220.

Golmohammadi, M., Shah, V., Lopez, S., Ziyabari, S., Yang, S., Camaratta, J., Obeid, I., and Picone, J. (2017a). The tuh eeg seizure corpus, in *ACNS Annual Meeting*, p. 1.

Golmohammadi, M., Torbati, A. H. H. N., de Diego, S. L., Obeid, I., and Picone, J. (2017b). Automatic analysis of eegs using big data and hybrid deep learning architectures, *arXiv preprint arXiv:1712.09771*.

Golmohammadi, M., Ziyabari, S., Shah, V., de Diego, S. L., Obeid, I., and Picone, J. (2017c). Deep architectures for automated seizure detection in scalp eegs, *arXiv preprint arXiv:1712.09776*.

Gong, C., Tao, D., Maybank, S. J., Liu, W., Kang, G., and Yang, J. (2016). Multi-modal curriculum learning for semi-supervised image classification, *IEEE Transactions on Image Processing* **25**, 7.

Goodfellow, I., Bengio, Y., and Courville, A. (2016). *Deep Learning* (MIT press).

Goodfellow, I., Pouget-Abadie, J., Mirza, M., Xu, B., Warde-Farley, D., Ozair, S., Courville, A., and Bengio, Y. (2014). Generative adversarial nets, in *Advances in Neural Information Processing Systems*, pp. 2672–2680.

Goodwin, T. R. and Harabagiu, S. M. (2017). Deep learning from eeg reports for inferring underspecified information, *AMIA Summits on Translational Science Proceedings* **2017**, pp. 112–121.

Gordienko, Y., Stirenko, S., Kochura, Y., Alienin, O., Novotarskiy, M., and Gordienko, N. (2017). Deep learning for fatigue estimation on the basis of multimodal human-machine interactions, *arXiv preprint arXiv:1801.06048*.

Gordon, S. M., Jaswa, M., Solon, A. J., and Lawhern, V. J. (2017). Real world bci: Cross-domain learning and practical applications, in *Proceedings of the 2017 ACM Workshop on An Application-Oriented Approach to BCI Out of the Laboratory* (ACM), pp. 25–28.

Guger, C., Daban, S., Sellers, E., Holzner, C., Krausz, G., Carabalona, R., Gramatica, F., and Edlinger, G. (2009). How many people are able to control a p300-based brain–computer interface (bci)? *Neuroscience Letters* **462**, 1, pp. 94–98.

Gui, Q., Jin, Z., and Xu, W. (2014). Exploring eeg-based biometrics for user identification and authentication, in *Signal Processing in Medicine and Biology Symposium (SPMB), 2014 IEEE* (IEEE), pp. 1–6.

Hachem, A., Khelifa, M. M. B., Alimi, A. M., Gorce, P., Arasu, S. V., Baulkani, S., Bisoy, S. K., Pattnaik, P. K., Ravindran, S., Palanisamy, N. *et al.* (2014).

Effect of fatigue on ssvep during virtual wheelchair navigation, *Journal of Theoretical and Applied Information Technology* **65**, 1.

Haider, A. and Fazel-Rezai, R. (2017). Application of p300 event-related potential in brain-computer interface, in *Event-Related Potentials and Evoked Potentials* (InTech).

Hajinoroozi, M., Jung, T.-P., Lin, C.-T., and Huang, Y. (2015a). Feature extraction with deep belief networks for driver's cognitive states prediction from eeg data, in *Signal and Information Processing (ChinaSIP), 2015 IEEE China Summit and International Conference on* (IEEE), pp. 812–815.

Hajinoroozi, M., Mao, Z., and Huang, Y. (2015b). Prediction of driver's drowsy and alert states from eeg signals with deep learning, in *Computational Advances in Multi-Sensor Adaptive Processing (CAMSAP), 2015 IEEE 6th International Workshop on* (IEEE), pp. 493–496.

Hajinoroozi, M., Mao, Z., Lin, Y.-P., and Huang, Y. (2017). Deep transfer learning for cross-subject and cross-experiment prediction of image rapid serial visual presentation events from eeg data, in *International Conference on Augmented Cognition* (Springer), pp. 45–55.

Hammond, D. C. (2003). The effects of caffeine on the brain: A review, *Journal of Neurotherapy* **7**, 2, pp. 79–89.

Han, C., Hayashi, H., Rundo, L., Araki, R., Shimoda, W., Muramatsu, S., Furukawa, Y., Mauri, G., and Nakayama, H. (2018). Gan-based synthetic brain mr image generation, in *2018 IEEE 15th International Symposium on Biomedical Imaging (ISBI 2018)* (IEEE), pp. 734–738.

Harati, A., Golmohammadi, M., Lopez, S., Obeid, I., and Picone, J. (2015). Improved eeg event classification using differential energy, in *Signal Processing in Medicine and Biology Symposium (SPMB), 2015 IEEE* (IEEE), pp. 1–4.

Harmony, T. (2013). The functional significance of delta oscillations in cognitive processing, *Frontiers in Integrative Neuroscience* **7**, p. 83.

Hartmann, K. G., Schirrmeister, R. T., and Ball, T. (2018). Hierarchical internal representation of spectral features in deep convolutional networks trained for eeg decoding, in *Brain-Computer Interface (BCI), 2018 6th International Conference on* (IEEE), pp. 1–6.

Hasasneh, A., Kampel, N., Sripad, P., Shah, N. J., and Dammers, J. (2018). Deep learning approach for automatic classification of ocular and cardiac artifacts in meg data, *Journal of Engineering* **2018**.

Havaei, M., Davy, A., Warde-Farley, D., Biard, A., Courville, A., Bengio, Y., Pal, C., Jodoin, P.-M., and Larochelle, H. (2017). Brain tumor segmentation with deep neural networks, *Medical Image Analysis* **35**, pp. 18–31.

Hennrich, J., Herff, C., Heger, D., and Schultz, T. (2015). Investigating deep learning for fnirs based bci. in *EMBC*, pp. 2844–2847.

Hernández, L. G., Mozos, O. M., Ferrández, J. M., and Antelis, J. M. (2018). Eeg-based detection of braking intention under different car driving conditions, *Frontiers in Neuroinformatics* **12**.

Hinton, G., Srivastava, N., and Swersky, K. (2012). Rmsprop: Divide the gradient by a running average of its recent magnitude, *Neural Networks for Machine Learning, Coursera Lecture 6e*.

Hinton, G. E. and Salakhutdinov, R. R. (2006). Reducing the dimensionality of data with neural networks, *Science* **313**, 5786, pp. 504–507.

Hiroyasu, T., Hanawa, K., and Yamamoto, U. (2014). Gender classification of subjects from cerebral blood flow changes using deep learning, in *Computational Intelligence and Data Mining (CIDM), 2014 IEEE Symposium on* (IEEE), pp. 229–233.

Hoang, T. and Choi, D. (2014). Secure and privacy enhanced gait authentication on smart phone, *The Scientific World Journal* **2014**.

Homer, M. L., Nurmikko, A. V., Donoghue, J. P., and Hochberg, L. R. (2013). Sensors and decoding for intracortical brain computer interfaces, *Annual Review of Biomedical Engineering* **15**, pp. 383–405.

Hosseini, M.-P., Pompili, D., Elisevich, K., and Soltanian-Zadeh, H. (2017a). Optimized deep learning for eeg big data and seizure prediction bci via internet of things, *IEEE Transactions on Big Data* **3**, 4, pp. 392–404.

Hosseini, M.-P., Soltanian-Zadeh, H., Elisevich, K., and Pompili, D. (2017b). Cloud-based deep learning of big eeg data for epileptic seizure prediction, *arXiv preprint arXiv:1702.05192*.

Hosseini, M.-P., Tran, T. X., Pompili, D., Elisevich, K., and Soltanian-Zadeh, H. (2017c). Deep learning with edge computing for localization of epileptogenicity using multimodal rs-fmri and eeg big data, in *Autonomic Computing (ICAC), 2017 IEEE International Conference on* (IEEE), pp. 83–92.

Hu, C., Ju, R., Shen, Y., Zhou, P., and Li, Q. (2016). Clinical decision support for alzheimer's disease based on deep learning and brain network, in *Communications (ICC), 2016 IEEE International Conference on* (IEEE), pp. 1–6.

Huang, D., Qian, K., Fei, D.-Y., Jia, W., Chen, X., and Bai, O. (2012). Electroencephalography (eeg)-based brain–computer interface (bci): A 2-d virtual wheelchair control based on event-related desynchronization/synchronization and state control, *IEEE Transactions on Neural Systems and Rehabilitation Engineering* **20**, 3, pp. 379–388.

Hung, Y.-C., Wang, Y.-K., Prasad, M., and Lin, C.-T. (2017). Brain dynamic states analysis based on 3d convolutional neural network, in *Systems, Man, and Cybernetics (SMC), 2017 IEEE International Conference on* (IEEE), pp. 222–227.

Huve, G., Takahashi, K., and Hashimoto, M. (2017). Brain activity recognition with a wearable fnirs using neural networks, in *Mechatronics and Automation (ICMA), 2017 IEEE International Conference on* (IEEE), pp. 1573–1578.

Huve, G., Takahashi, K., and Hashimoto, M. (2018). Brain-computer interface using deep neural network and its application to mobile robot control, in *Advanced Motion Control (AMC), 2018 IEEE 15th International Workshop on* (IEEE), pp. 169–174.

Jayarathne, I., Cohen, M., and Amarakeerthi, S. (2016). Brainid: Development of an eeg-based biometric authentication system, in *Information Technology, Electronics and Mobile Communication Conference (IEMCON), 2016 IEEE 7th Annual* (IEEE), pp. 1–6.

Jia, X., Li, K., Li, X., and Zhang, A. (2014). A novel semi-supervised deep learning framework for affective state recognition on eeg signals, in *Bioinformatics and Bioengineering (BIBE), 2014 IEEE International Conference on* (IEEE), pp. 30–37.

Jingwei, L., Yin, C., and Weidong, Z. (2015). Deep learning eeg response representation for brain computer interface, in *Control Conference (CCC), 2015 34th Chinese* (IEEE), pp. 3518–3523.

Johansen, A. R., Jin, J., Maszczyk, T., Dauwels, J., Cash, S. S., and Westover, M. B. (2016). Epileptiform spike detection via convolutional neural networks, in *Acoustics, Speech and Signal Processing (ICASSP), 2016 IEEE International Conference on* (IEEE), pp. 754–758.

Joshi, R., Goel, P., Sur, M., and Murthy, H. A. (2018). Single trial p300 classification using convolutional lstm and deep learning ensembles method, in *International Conference on Intelligent Human Computer Interaction* (Springer), pp. 3–15.

Kamburugamuve, S., Christiansen, L., and Fox, G. (2015). A framework for real time processing of sensor data in the cloud, *Journal of Sensors* **2015**.

Kang, H. and Choi, S. (2014). Bayesian common spatial patterns for multi-subject eeg classification, *Neural Networks* **57**, pp. 39–50.

Kanoga, S., Hoshino, T., and Asoh, H. (2020). Independent low-rank matrix analysis-based automatic artifact reduction technique applied to three bci paradigms, *Frontiers in Human Neuroscience*.

Kavasidis, I., Palazzo, S., Spampinato, C., Giordano, D., and Shah, M. (2017). Brain2image: Converting brain signals into images, in *Proceedings of the 25th ACM International Conference on Multimedia* (ACM), pp. 1809–1817.

Kawasaki, K., Yoshikawa, T., and Furuhashi, T. (2015). Visualizing extracted feature by deep learning in p300 discrimination task, in *Soft Computing and Pattern Recognition (SoCPaR), 2015 7th International Conference of* (IEEE), pp. 149–154.

Kawde, P. and Verma, G. K. (2017). Deep belief network based affect recognition from physiological signals, in *Electrical, Computer and Electronics (UPCON), 2017 4th IEEE Uttar Pradesh Section International Conference on* (IEEE), pp. 587–592.

Kerem, D. H. and Geva, A. B. (2017). Brain state identification and forecasting of acute pathology using unsupervised fuzzy clustering of eeg temporal patterns, in *Fuzzy and Neuro-Fuzzy Systems in Medicine* (CRC Press), pp. 19–68.

Keshishzadeh, S., Fallah, A., and Rashidi, S. (2016). Improved eeg based human authentication system on large dataset, in *Electrical Engineering (ICEE), 2016 24th Iranian Conference on* (IEEE), pp. 1165–1169.

Khurana, P., Majumdar, A., and Ward, R. (2016). Class-wise deep dictionaries for eeg classification, in *Neural Networks (IJCNN), 2016 International Joint Conference on* (IEEE), pp. 3556–3563.

Kidmose, P., Looney, D., Ungstrup, M., Rank, M. L., and Mandic, D. P. (2013). A study of evoked potentials from ear-eeg, *IEEE Transactions on Biomedical Engineering* **60**, 10, pp. 2824–2830.

Kim, Y., Ryu, J., Kim, K. K., Took, C. C., Mandic, D. P., and Park, C. (2016). Motor imagery classification using mu and beta rhythms of eeg with strong uncorrelating transform based complex common spatial patterns, *Computational Intelligence and Neuroscience* **2016**, pp. 1–14.

Kingma, D. and Ba, J. (2014). Adam: A method for stochastic optimization, *arXiv preprint arXiv:1412.6980*.

Kingma, D. P., Mohamed, S., Rezende, D. J., and Welling, M. (2014). Semi-supervised learning with deep generative models, in *Advances in Neural Information Processing Systems (NIPS)*.

Kingma, D. P. and Welling, M. (2013). Auto-encoding variational bayes, *arXiv preprint arXiv:1312.6114*.

Kipf, T. N. and Welling, M. (2017). Semi-supervised classification with graph convolutional networks, *ICLR*.

Kiral-Kornek, I., Roy, S., Nurse, E., Mashford, B., Karoly, P., Carroll, T., Payne, D., Saha, S., Baldassano, S., O'Brien, T. *et al.* (2018). Epileptic seizure prediction using big data and deep learning: Toward a mobile system, *EBioMedicine* **27**, pp. 103–111.

Knott, V., Cosgrove, M., Villeneuve, C., Fisher, D., Millar, A., and McIntosh, J. (2008). Eeg correlates of imagery-induced cigarette craving in male and female smokers, *Addictive Behaviors* **33**, 4, pp. 616–621.

Knyazev, G. G. (2012). Eeg delta oscillations as a correlate of basic homeostatic and motivational processes, *Neuroscience & Biobehavioral Reviews* **36**, 1, pp. 677–695.

Koike-Akino, T., Mahajan, R., Marks, T. K., Wang, Y., Watanabe, S., Tuzel, O., and Orlik, P. (2016). High-accuracy user identification using eeg biometrics, in *2016 38th Annual International Conference of the IEEE Engineering in Medicine and Biology Society (EMBC)* (IEEE), pp. 854–858.

Konno, S., Nakamura, Y., Shiraishi, Y., and Takahashi, O. (2015). Gait-based authentication using trouser front-pocket sensors.

Koyamada, S., Shikauchi, Y., Nakae, K., Koyama, M., and Ishii, S. (2015). Deep learning of fmri big data: A novel approach to subject-transfer decoding, *arXiv preprint arXiv:1502.00093*.

Krizhevsky, A., Sutskever, I., and Hinton, G. E. (2012). Imagenet classification with deep convolutional neural networks, in *NeurIPS*, pp. 1097–1105.

Kulasingham, J., Vibujithan, V., and De Silva, A. (2016). Deep belief networks and stacked autoencoders for the p300 guilty knowledge test, in *Biomedical Engineering and Sciences (IECBES), 2016 IEEE EMBS Conference on* (IEEE), pp. 127–132.

Kumar, S., Sharma, A., Mamun, K., and Tsunoda, T. (2016). A deep learning approach for motor imagery eeg signal classification, in *Computer Science and Engineering (APWC on CSE), 2016 3rd Asia-Pacific World Congress on* (IEEE), pp. 34–39.

Kwak, N.-S., Müller, K.-R., and Lee, S.-W. (2017). A convolutional neural network for steady state visual evoked potential classification under ambulatory environment, *PloS One* **12**, 2, p. e0172578.

Latman, N. S. and Herb, E. (2013). A field study of the accuracy and reliability of a biometric iris recognition system, *Science & Justice* **53**, 2, pp. 98–102.

Lawhern, V., Solon, A., Waytowich, N., Gordon, S. M., Hung, C., and Lance, B. J. (2018). Eegnet: A compact convolutional neural network for eeg-based brain–computer interfaces, *Journal of Neural Engineering*.

Lee, H. K. and Choi, Y.-S. (2018). A convolution neural networks scheme for classification of motor imagery eeg based on wavelet time-frequecy image, in *Information Networking (ICOIN), 2018 International Conference on* (IEEE), pp. 906–909.

Lees, S., Dayan, N., Cecotti, H., Mccullagh, P., Maguire, L., Lotte, F., and Coyle, D. (2018). A review of rapid serial visual presentation-based brain–computer interfaces, *Journal of Neural Engineering* **15**, 2, p. 021001.

Leuthardt, E. C., Schalk, G., Roland, J., Rouse, A., and Moran, D. W. (2009). Evolution of brain-computer interfaces: Going beyond classic motor physiology, *Neurosurgical Focus* **27**, 1, p. E4.

Li, J. and Cichocki, A. (2014). Deep learning of multifractal attributes from motor imagery induced eeg, in *International Conference on Neural Information Processing* (Springer), pp. 503–510.

Li, J., Struzik, Z., Zhang, L., and Cichocki, A. (2015a). Feature learning from incomplete eeg with denoising autoencoder, *Neurocomputing* **165**, pp. 23–31.

Li, J., Zhang, Z., and He, H. (2016a). Implementation of eeg emotion recognition system based on hierarchical convolutional neural networks, in *International Conference on Brain Inspired Cognitive Systems* (Springer), pp. 22–33.

Li, J., Zhang, Z., and He, H. (2017a). Hierarchical convolutional neural networks for eeg-based emotion recognition, *Cognitive Computation*, pp. 1–13.

Li, K., Li, X., Zhang, Y., and Zhang, A. (2013). Affective state recognition from eeg with deep belief networks, in *2013 IEEE International Conference on Bioinformatics and Biomedicine* (IEEE), pp. 305–310.

Li, P., Jiang, W., and Su, F. (2016b). Single-channel eeg-based mental fatigue detection based on deep belief network, in *Engineering in Medicine and Biology Society (EMBC), 2016 IEEE 38th Annual International Conference of the* (IEEE), pp. 367–370.

Li, P., Peng, L., Cai, J., Ding, X., and Ge, S. (2017b). Attention based rnn model for document image quality assessment, in *Document Analysis and Recognition (ICDAR), 2017 14th IAPR International Conference on*, Vol. 1 (IEEE), pp. 819–825.

Li, R., Zhang, W., Suk, H.-I., Wang, L., Li, J., Shen, D., and Ji, S. (2014). Deep learning based imaging data completion for improved brain disease diagnosis, in *International Conference on Medical Image Computing and Computer-Assisted Intervention* (Springer), pp. 305–312.

Li, X. and Duncan, J. (2020). Braingnn: Interpretable brain graph neural network for fmri analysis, *bioRxiv*.

Li, X., Qian, B., Wei, J., Li, A., Liu, X., and Zheng, Q. (2019). Classify eeg and reveal latent graph structure with spatio-temporal graph convolutional neural network, in *2019 IEEE International Conference on Data Mining (ICDM)* (IEEE), pp. 389–398.

Li, X., Zhang, P., Song, D., Yu, G., Hou, Y., and Hu, B. (2015b). Eeg based emotion identification using unsupervised deep feature learning.

Li, Y., Guan, C., Li, H., and Chin, Z. (2008). A self-training semi-supervised svm algorithm and its application in an eeg-based brain computer interface speller system, *Pattern Recognition Letters* **29**, 9, pp. 1285–1294.

Lin, Q., Ye, S.-q., Huang, X.-m., Li, S.-y., Zhang, M.-z., Xue, Y., and Chen, W.-S. (2016). Classification of epileptic eeg signals with stacked sparse autoencoder based on deep learning, in *International Conference on Intelligent Computing* (Springer), pp. 802–810.

Lin, Z., Zeng, Y., Tong, L., Zhang, H., Zhang, C., and Yan, B. (2017). Method for enhancing single-trial p300 detection by introducing the complexity degree of image information in rapid serial visual presentation tasks, *PloS One* **12**, 12, p. e0184713.

Lipton, Z. C., Kale, D. C., Elkan, C., and Wetzel, R. (2015). Learning to diagnose with lstm recurrent neural networks, *arXiv preprint arXiv:1511.03677*.

Liu, J., Pan, Y., Li, M., Chen, Z., Tang, L., Lu, C., and Wang, J. (2018a). Applications of deep learning to mri images: A survey, *Big Data Mining and Analytics* **1**, 1, pp. 1–18.

Liu, M., Wu, W., Gu, Z., Yu, Z., Qi, F., and Li, Y. (2018b). Deep learning based on batch normalization for p300 signal detection, *Neurocomputing* **275**, pp. 288–297.

Liu, Q., Zhao, X.-G., Hou, Z.-G., and Liu, H.-G. (2017a). Deep belief networks for eeg-based concealed information test, in *International Symposium on Neural Networks* (Springer), pp. 498–506.

Liu, W., Jiang, H., and Lu, Y. (2017b). Analyze eeg signals with convolutional neural network based on power spectrum feature selection, *Proceedings of Science*.

Liu, W., Zheng, W.-L., and Lu, B.-L. (2016). Emotion recognition using multimodal deep learning, in *International Conference on Neural Information Processing* (Springer), pp. 521–529.

Long, T., Thai, L., and Hanh, T. (2012). Multimodal biometric person authentication using fingerprint, face features, *PRICAI 2012: Trends in Artificial Intelligence*, pp. 613–624.

Lotte, F., Bougrain, L., Cichocki, A., Clerc, M., Congedo, M., Rakotomamonjy, A., and Yger, F. (2018). A review of classification algorithms for eeg-based brain–computer interfaces: A 10 year update, *Journal of Neural Engineering* **15**, 3, p. 031005.

Lotte, F., Congedo, M., Lécuyer, A., Lamarche, F., and Arnaldi, B. (2007). A review of classification algorithms for eeg-based brain–computer interfaces, *Journal of Neural Engineering* **4**, 2, p. R1.

Lu, N., Li, T., Ren, X., and Miao, H. (2017). A deep learning scheme for motor imagery classification based on restricted boltzmann machines,

IEEE *Transactions on Neural Systems and Rehabilitation Engineering* **25**, 6, pp. 566–576.

Luong, M.-T., Pham, H., and Manning, C. D. (2015). Effective approaches to attention-based neural machine translation, *arXiv preprint arXiv: 1508.04025*.

Ma, L., Minett, J. W., Blu, T., and Wang, W. S. (2015). Resting state eeg-based biometrics for individual identification using convolutional neural networks, in *EMBC* (IEEE), pp. 2848–2851.

Ma, T., Li, H., Yang, H., Lv, X., Li, P., Liu, T., Yao, D., and Xu, P. (2017). The extraction of motion-onset vep bci features based on deep learning and compressed sensing, *Journal of Neuroscience Methods* **275**, pp. 80–92.

Ma, X., Qiu, S., Du, C., Xing, J., and He, H. (2018). Improving eeg-based motor imagery classification via spatial and temporal recurrent neural networks, in *EMBC* (IEEE), pp. 1903–1906.

Maddula, R., Stivers, J., Mousavi, M., Ravindran, S., and de Sa, V. (2017). Deep recurrent convolutional neural networks for classifying p300 bci signals, in *Proceedings of the 7th Graz Brain-Computer Interface Conference, Graz, Austria*, pp. 18–22.

Mahmud, M., Kaiser, M. S., Hussain, A., and Vassanelli, S. (2018). Applications of deep learning and reinforcement learning to biological data, *IEEE Transactions on Neural Networks and Learning Systems* **29**, 6, pp. 2063–2079.

Makhzani, A., Shlens, J., Jaitly, N., Goodfellow, I., and Frey, B. (2015). Adversarial autoencoders, *arXiv preprint arXiv:1511.05644*.

Manjunathswamy, B., Abhishek, A. M., Thriveni, J., Venugopal, K., and Patnaik, L. (2015). Multimodal biometric authentication using ecg and fingerprint, *International Journal of Computer Applications* **111**, 13.

Manor, R. and Geva, A. B. (2015). Convolutional neural network for multi-category rapid serial visual presentation bci, *Frontiers in Computational Neuroscience* **9**, p. 146.

Manor, R., Mishali, L., and Geva, A. B. (2016). Multimodal neural network for rapid serial visual presentation brain computer interface, *Frontiers in Computational Neuroscience* **10**, p. 130.

Manzano, M., Guillén, A., Rojas, I., and Herrera, L. J. (2017). Combination of eeg data time and frequency representations in deep networks for sleep stage classification, in *International Conference on Intelligent Computing* (Springer), pp. 219–229.

Mao, Z. (2016). *Deep Learning for Rapid Serial Visual Presentation Event from Electroencephalography Signal*, Ph.D. thesis, The University of Texas at San Antonio.

Mao, Z., Lawhern, V., Merino, L. M., Ball, K., Deng, L., Lance, B. J., Robbins, K., and Huang, Y. (2014). Classification of non-time-locked rapid serial visual presentation events for brain-computer interaction using deep learning, in *Signal and Information Processing (ChinaSIP), 2014 IEEE China Summit & International Conference on* (IEEE), pp. 520–524.

Mao, Z., Yao, W. X., and Huang, Y. (2017). Eeg-based biometric identification with deep learning, in *Neural Engineering (NER), 2017 8th International IEEE/EMBS Conference on* (IEEE), pp. 609–612.

Marc Moreno, L. (2017). Deep learning for brain tumor segmentation, *Master diss. University of Colorado Colorado Springs.*

Markham, B. and Townshend, J. (1981). Land cover classification accuracy as a function of sensor spatial resolution.

McGinty, D., Szymusiak, R., and Thomson, D. (1994). Preoptic/anterior hypothalamic warming increases eeg delta frequency activity within non-rapid eye movement sleep, *Brain Research* **667**, 2, pp. 273–277.

Meisheri, H., Ramrao, N., and Mitra, S. (2018). Multiclass common spatial pattern for eeg based brain computer interface with adaptive learning classifier, *arXiv preprint arXiv:1802.09046.*

Mikkelsen, K. B., Kappel, S. L., Mandic, D. P., and Kidmose, P. (2015). Eeg recorded from the ear: Characterizing the ear-eeg method, *Frontiers in Neuroscience* **9**, p. 438.

Mikolov, T., Karafiát, M., Burget, L., Cernockỳ, J., and Khudanpur, S. (2010). Recurrent neural network based language model. in *Interspeech*, Vol. 2, p. 3.

Min, S., Lee, B., and Yoon, S. (2017). Deep learning in bioinformatics, *Briefings in Bioinformatics* **18**, 5, pp. 851–869.

Mioranda-Correa, J. A. and Patras, I. (2018). A multi-task cascaded network for prediction of affect, personality, mood and social context using eeg signals, in *Automatic Face & Gesture Recognition (FG 2018), 2018 13th IEEE International Conference on* (IEEE), pp. 373–380.

Mirza, M. and Osindero, S. (2014). Conditional generative adversarial nets, *arXiv preprint arXiv:1411.1784.*

Mnih, V., Kavukcuoglu, K., Silver, D., Rusu, A. A., Veness, J., Bellemare, M. G., Graves, A., Riedmiller, M., Fidjeland, A. K., Ostrovski, G. *et al.* (2015). Human-level control through deep reinforcement learning, *Nature* **518**, 7540, p. 529.

Morabito, F. C., Campolo, M., Ieracitano, C., Ebadi, J. M., Bonanno, L., Bramanti, A., Desalvo, S., Mammone, N., and Bramanti, P. (2016). Deep convolutional neural networks for classification of mild cognitive impaired and alzheimer's disease patients from scalp eeg recordings, in *Research and Technologies for Society and Industry Leveraging a Better Tomorrow (RTSI), 2016 IEEE 2nd International Forum on* (IEEE), pp. 1–6.

Morabito, F. C., Campolo, M., Mammone, N., Versaci, M., Franceschetti, S., Tagliavini, F., Sofia, V., Fatuzzo, D., Gambardella, A., Labate, A. *et al.* (2017). Deep learning representation from electroencephalography of early-stage creutzfeldt-jakob disease and features for differentiation from rapidly progressive dementia, *International Journal of Neural Systems* **27**, 02, p. 1650039.

Müller-Gerking, J., Pfurtscheller, G., and Flyvbjerg, H. (1999). Designing optimal spatial filters for single-trial eeg classification in a movement task, *Clinical Neurophysiology* **110**, 5, pp. 787–798.

Muramatsu, D., Shiraishi, A., Makihara, Y., Uddin, M. Z., and Yagi, Y. (2015). Gait-based person recognition using arbitrary view transformation model, *IEEE Transactions on Image Processing* **24**, 1, pp. 140–154.

Nagi, J., Ducatelle, F., Di Caro, G. A., Cireşan, D., Meier, U., Giusti, A., Nagi, F., Schmidhuber, J., and Gambardella, L. M. (2011). Max-pooling convolutional neural networks for vision-based hand gesture recognition, in *Signal and Image Processing Applications (ICSIPA)* (IEEE), pp. 342–347.

Narayanaswamy, S., Paige, T. B., Van de Meent, J.-W., Desmaison, A., Goodman, N., Kohli, P., Wood, F., and Torr, P. (2017). Learning disentangled representations with semi-supervised deep generative models, in *NIPS*.

Narejo, S., Pasero, E., and Kulsoom, F. (2016). Eeg based eye state classification using deep belief network and stacked autoencoder, *International Journal of Electrical and Computer Engineering (IJECE)* **6**, 6, pp. 3131–3141.

Naseer, N. and Hong, K.-S. (2015). fnirs-based brain-computer interfaces: A review, *Frontiers in Human Neuroscience* **9**, p. 3.

Naseer, N., Qureshi, N. K., Noori, F. M., and Hong, K.-S. (2016). Analysis of different classification techniques for two-class functional near-infrared spectroscopy-based brain-computer interface, *Computational Intelligence and Neuroscience* **2016**.

Nguyen, T., Nahavandi, S., Khosravi, A., Creighton, D., and Hettiarachchi, I. (2015). Eeg signal analysis for bci application using fuzzy system, in *Neural Networks (IJCNN), 2015 International Joint Conference on* (IEEE), pp. 1–8.

Nguyen, T.-H. and Chung, W.-Y. (2018). A single-channel ssvep-based bci speller using deep learning, *IEEE Access* **7**, pp. 1752–1763.

Ning, R., Wang, C., Xin, C., Li, J., and Wu, H. (2018). Deepmag: Sniffing mobile apps in magnetic field through deep convolutional neural networks, in *2018 IEEE International Conference on Pervasive Computing and Communications (PerCom)* (IEEE), pp. 1–10.

Nishimoto, S., Vu, A. T., Naselaris, T., Benjamini, Y., Yu, B., and Gallant, J. L. (2011). Reconstructing visual experiences from brain activity evoked by natural movies, *Current Biology* **21**, 19, pp. 1641–1646.

Norcia, A. M., Appelbaum, L. G., Ales, J. M., Cottereau, B. R., and Rossion, B. (2015). The steady-state visual evoked potential in vision research: A review, *Journal of Vision* **15**, 6, pp. 4–4.

Nurse, E., Mashford, B. S., Yepes, A. J., Kiral-Kornek, I., Harrer, S., and Freestone, D. R. (2016). Decoding eeg and lfp signals using deep learning: Heading truenorth, in *Proceedings of the ACM International Conference on Computing Frontiers* (ACM), pp. 259–266.

Nurse, E. S., Karoly, P. J., Grayden, D. B., and Freestone, D. R. (2015). A generalizable brain-computer interface (bci) using machine learning for feature discovery, *PloS One* **10**, 6, p. e0131328.

Obeid, I. and Picone, J. (2016). The temple university hospital eeg data corpus, *Frontiers in Neuroscience* **10**.

Odena, A. (2016). Semi-supervised learning with generative adversarial networks, *arXiv preprint arXiv:1606.01583*.

Odena, A., Olah, C., and Shlens, J. (2017). Conditional image synthesis with auxiliary classifier gans, in *Proceedings of the 34th International Conference on Machine Learning-Volume 70* (JMLR. org), pp. 2642–2651.

or Rashid, M. M. and Ahmad, M. (2016). Classification of motor imagery hands movement using levenberg-marquardt algorithm based on statistical features of eeg signal, in *ICEEICT* (IEEE), pp. 1–6.

Ormerod, D. (2017). Sounding out expert voice identification, *Expert Evidence and Scientific Proof in Criminal Trials*.

Orosco, L., Correa, A. G., Diez, P., and Laciar, E. (2016). Patient non-specific algorithm for seizures detection in scalp eeg, *Computers in Biology and Medicine* **71**, pp. 128–134.

Ortiz, A., Munilla, J., Gorriz, J. M., and Ramirez, J. (2016). Ensembles of deep learning architectures for the early diagnosis of the alzheimer's disease, *International Journal of Neural Systems* **26**, 07, p. 1650025.

Pacharra, M., Debener, S., and Wascher, E. (2017). Concealed around-the-ear eeg captures cognitive processing in a visual simon task, *Frontiers in Human Neuroscience* **11**, p. 290.

Page, A., Turner, J., Mohsenin, T., and Oates, T. (2014). Comparing raw data and feature extraction for seizure detection with deep learning methods. in *FLAIRS Conference*.

Palazzo, S., Spampinato, C., Kavasidis, I., Giordano, D., and Shah, M. (2017). Generative adversarial networks conditioned by brain signals, in *Proceedings of the IEEE International Conference on Computer Vision*, pp. 3410–3418.

Pandarinath, C., Nuyujukian, P., Blabe, C. H., Sorice, B. L., Saab, J., Willett, F. R., Hochberg, L. R., Shenoy, K. V., and Henderson, J. M. (2017). High performance communication by people with paralysis using an intracortical brain-computer interface, *Elife* **6**, p. e18554.

Parasuraman, R. and Jiang, Y. (2012). Individual differences in cognition, affect, and performance: Behavioral, neuroimaging, and molecular genetic approaches, *Neuroimage* **59**, 1, pp. 70–82.

Parra, L. C., Spence, C. D., Gerson, A. D., and Sajda, P. (2005). Recipes for the linear analysis of eeg, *Neuroimage* **28**, 2, pp. 326–341.

Pasqualetti, F., Dörfler, F., and Bullo, F. (2013). Attack detection and identification in cyber-physical systems, *IEEE Transactions on Automatic Control* **58**, 11, pp. 2715–2729.

Peng, L., Chen, W., Zhou, W., Li, F., Yang, J., and Zhang, J. (2016). An immune-inspired semi-supervised algorithm for breast cancer diagnosis, *Computer Methods and Programs in Biomedicine* **134**.

Pereira, A., Padden, D., Jantz, J., Lin, K., and Alcaide-Aguirre, R. (2018). Cross-subject eeg event-related potential classification for brain-computer interfaces using residual networks.

Pérez-Benítez, J., Pérez-Benítez, J., and Espina-Hernández, J. (2018). Development of a brain computer interface interface using multi-frequency visual stimulation and deep neural networks, in *Electronics, Communications and Computers (CONIELECOMP), 2018 International Conference on* (IEEE), pp. 18–24.

Pfurtscheller, G. and Da Silva, F. L. (1999). Event-related eeg/meg synchronization and desynchronization: Basic principles, *Clinical Neurophysiology* **110**, 11, pp. 1842–1857.

Pfurtscheller, G. and Neuper, C. (2001). Motor imagery and direct brain-computer communication, *Proceedings of the IEEE* **89**, 7, pp. 1123–1134.

Pinheiro, O. R., Alves, L. R., Romero, M., and de Souza, J. R. (2016). Wheelchair simulator game for training people with severe disabilities, in *Technology and Innovation in Sports, Health and Wellbeing (TISHW), International Conference on* (IEEE).

Plis, S. M., Hjelm, D. R., Salakhutdinov, R., Allen, E. A., Bockholt, H. J., Long, J. D., Johnson, H. J., Paulsen, J. S., Turner, J. A., and Calhoun, V. D. (2014). Deep learning for neuroimaging: A validation study, *Frontiers in Neuroscience* **8**, p. 229.

Putten, M. J., Olbrich, S., and Arns, M. (2018). Predicting sex from brain rhythms with deep learning, *Scientific Reports* **8**, 1, p. 3069.

Qian, K., Wu, C., Yang, Z., Zhou, Z., Wang, X., and Liu, Y. (2018). Enabling phased array signal processing for mobile wifi devices, *IEEE Transactions on Mobile Computing* **17**, 8, pp. 1820–1833.

Radford, A., Metz, L., and Chintala, S. (2016). Unsupervised representation learning with deep convolutional generative adversarial networks, *International Conference on Learning Representations (ICLR)*.

Ramoser, H., Muller-Gerking, J., and Pfurtscheller, G. (2000). Optimal spatial filtering of single trial eeg during imagined hand movement, *IEEE Transactions on Rehabilitation Engineering* **8**, 4, pp. 441–446.

Ravanelli, M. and Bengio, Y. (2018). Speaker recognition from raw waveform with sincnet, in *2018 IEEE Spoken Language Technology Workshop (SLT)* (IEEE), pp. 1021–1028.

Ravi, A., Beni, N. H., Manuel, J., and Jiang, N. (2020). Comparing user-dependent and user-independent training of cnn for ssvep bci, *Journal of Neural Engineering* **17**, 2, p. 026028.

Reddy, T. K. and Behera, L. (2016). Online eye state recognition from eeg data using deep architectures, in *Systems, Man, and Cybernetics (SMC), 2016 IEEE International Conference on* (IEEE), pp. 000712–000717.

Redkar, S. (2015). Using deep learning for human computer interface via electroencephalography, *IAES International Journal of Robotics and Automation* **4**, 4.

Regan, D. (1977). Steady-state evoked potentials, *JOSA* **67**, 11, pp. 1475–1489.

Reid, M. S., Flammino, F., Howard, B., Nilsen, D., and Prichep, L. S. (2006). Topographic imaging of quantitative eeg in response to smoked cocaine self-administration in humans, *Neuropsychopharmacology* **31**, 4, p. 872.

Ren, S., He, K., Girshick, R., and Sun, J. (2017). Faster r-cnn: Towards real-time object detection with region proposal networks, *IEEE Transactions on Pattern Analysis & Machine Intelligence*, 6, pp. 1137–1149.

Ren, Y. and Wu, Y. (2014). Convolutional deep belief networks for feature extraction of eeg signal, in *Neural Networks (IJCNN), 2014 International Joint Conference on* (IEEE), pp. 2850–2853.

Rodrigues, D., Silva, G. F., Papa, J. P., Marana, A. N., and Yang, X.-S. (2016). Eeg-based person identification through binary flower pollination algorithm, *Expert Systems with Applications* **62**, pp. 81–90.

Roy, Y., Banville, H., Albuquerque, I., Gramfort, A., Falk, T. H., and Faubert, J. (2019). Deep learning-based electroencephalography analysis: A systematic review, *Journal of Neural Engineering* **16**, 5, p. 051001.

Ruffini, G., Ibañez, D., Castellano, M., Dunne, S., and Soria-Frisch, A. (2016). Eeg-driven rnn classification for prognosis of neurodegeneration in at-risk patients, in *International Conference on Artificial Neural Networks* (Springer), pp. 306–313.

Russoniello, C. V., O'Brien, K., and Parks, J. M. (2009). The effectiveness of casual video games in improving mood and decreasing stress, *Journal of CyberTherapy & Rehabilitation* **2**, 1, pp. 53–66.

Rusydi, M., Okamoto, T., Ito, S., and Sasaki, M. (2014). Rotation matrix to operate a robot manipulator for 2d analog tracking objects using electrooculography, *Robotics* **3**, 3, pp. 289–309.

Sachdev, R. N., Gaspard, N., Gerrard, J. L., Hirsch, L. J., Spencer, D. D., and Zaveri, H. P. (2015). Delta rhythm in wakefulness: Evidence from intracranial recordings in human beings, *Journal of Neurophysiology* **114**, 2, pp. 1248–1254.

Sadikoglu, F. and Uzelaltinbulat, S. (2016). Biometric retina identification based on neural network, *Procedia Computer Science* **102**, pp. 26–33.

Saha, S. and Baumert, M. (2019). Intra-and inter-subject variability in eeg-based sensorimotor brain computer interface: A review, *Frontiers in Computational Neuroscience* **13**, p. 87.

Sak, H., Senior, A., and Beaufays, F. (2014). Long short-term memory recurrent neural network architectures for large scale acoustic modeling, in *Fifteenth Annual Conference of the International Speech Communication Association*.

Sakhavi, S., Guan, C., and Yan, S. (2015). Parallel convolutional-linear neural network for motor imagery classification, in *Signal Processing Conference (EUSIPCO), 2015 23rd European* (IEEE), pp. 2736–2740.

Salimans, T., Goodfellow, I., Zaremba, W., Cheung, V., Radford, A., and Chen, X. (2016). Improved techniques for training gans, in *Advances in Neural Information Processing Systems (NIPS)*.

Samek, W., Müller, K.-R., Kawanabe, M., and Vidaurre, C. (2012). Brain-computer interfacing in discriminative and stationary subspaces, in *Engineering in Medicine and Biology Society (EMBC), 2012 Annual International Conference of the IEEE* (IEEE), pp. 2873–2876.

San, P. P., Ling, S. H., Chai, R., Tran, Y., Craig, A., and Nguyen, H. (2016). Eeg-based driver fatigue detection using hybrid deep generic model, in *Engineering in Medicine and Biology Society (EMBC), 2016 IEEE 38th Annual International Conference of the* (IEEE), pp. 800–803.

Sanz-Martin, A., Guevara, M. Á., Amezcua, C., Santana, G., and Hernández-González, M. (2011). Effects of red wine on the electrical activity and functional coupling between prefrontal–parietal cortices in young men, *Appetite* **57**, 1, pp. 84–93.

Sarkar, S., Reddy, K., Dorgan, A., Fidopiastis, C., and Giering, M. (2016). Wearable eeg-based activity recognition in phm-related service environment via deep learning, *Int. J. Progn. Health Manag.* **7**, pp. 1–10.

Sarraf, S. and Tofighi, G. (2016). Deep learning-based pipeline to recognize alzheimer's disease using fmri data, in *Future Technologies Conference (FTC)* (IEEE), pp. 816–820.

Sarraf, S., Tofighi, G. *et al.* (2016). Deepad: Alzheimer disease classification via deep convolutional neural networks using mri and fmri, *bioRxiv*, p. 070441.

Sazgar, M. and Young, M. G. (2019). Overview of eeg, electrode placement, and montages, in *Absolute Epilepsy and EEG Rotation Review* (Springer), pp. 117–125.

Schalk, G., McFarland, D. J., Hinterberger, T., Birbaumer, N., and Wolpaw, J. R. (2004). Bci2000: A general-purpose brain-computer interface (bci) system, *IEEE Transactions on Biomedical Engineering* **51**, 6, pp. 1034–1043.

Schetinin, V., Jakaite, L., Nyah, N., Novakovic, D., and Krzanowski, W. (2017). Feature extraction with gmdh-type neural networks for eeg-based person identification, *IJNS*, p. 1750064.

Schirrmeister, R., Gemein, L., Eggensperger, K., Hutter, F., and Ball, T. (2017). Deep learning with convolutional neural networks for decoding and visualization of eeg pathology, in *Signal Processing in Medicine and Biology Symposium (SPMB), 2017 IEEE* (IEEE), pp. 1–7.

Seeliger, K., Güçlü, U., Ambrogioni, L., Güçlütürk, Y., and Van Gerven, M. (2018). Generative adversarial networks for reconstructing natural images from brain activity, *NeuroImage* **181**, pp. 775–785.

Shah, V., Golmohammadi, M., Ziyabari, S., Von Weltin, E., Obeid, I., and Picone, J. (2017). Optimizing channel selection for seizure detection, in *Signal Processing in Medicine and Biology Symposium (SPMB), 2017 IEEE* (IEEE), pp. 1–5.

Shahin, M., Ahmed, B., Hamida, S. T.-B., Mulaffer, F. L., Glos, M., and Penzel, T. (2017). Deep learning and insomnia: Assisting clinicians with their diagnosis, *IEEE Journal of Biomedical and Health Informatics* **21**, 6, pp. 1546–1553.

Shamwell, J., Lee, H., Kwon, H., Marathe, A. R., Lawhern, V., and Nothwang, W. (2016). Single-trial eeg rsvp classification using convolutional neural networks, in *Micro-and Nanotechnology Sensors, Systems, and Applications VIII*, Vol. 9836 (International Society for Optics and Photonics), p. 983622.

Shanbhag, A., Kholkar, A. P., Sawant, S., Vicente, A., Martires, S., and Patil, S. (2017). P300 analysis using deep neural network, in *2017 International Conference on Energy, Communication, Data Analytics and Soft Computing (ICECDS)* (IEEE), pp. 3142–3147.

Shang, J., Zhang, W., Xiong, J., and Liu, Q. (2017). Cognitive load recognition using multi-channel complex network method, in *International Symposium on Neural Networks* (Springer), pp. 466–474.

Shen, G., Horikawa, T., Majima, K., and Kamitani, Y. (2019). Deep image reconstruction from human brain activity, *PLoS Computational Biology* **15**, 1, p. e1006633.

Shenoy, H. V., Vinod, A. P., and Guan, C. (2015). Shrinkage estimator based regularization for eeg motor imagery classification, in *2015 10th International Conference on Information, Communications and Signal Processing (ICICS)* (IEEE), pp. 1–5.

Shreyas, V. and Pankajakshan, V. (2017). A deep learning architecture for brain tumor segmentation in mri images, in *Multimedia Signal Processing (MMSP), 2017 IEEE 19th International Workshop on* (IEEE), pp. 1–6.

Shu, M. and Fyshe, A. (2013). Sparse autoencoders for word decoding from magnetoencephalography, in *Proceedings of the third NIPS Workshop on Machine Learning and Interpretation in NeuroImaging (MLINI)*.

Sita, J. and Nair, G. (2013). Feature extraction and classification of eeg signals for mapping motor area of the brain, in *ICCC* (IEEE), pp. 463–468.

Snoek, J., Larochelle, H., and Adams, R. P. (2012). Practical bayesian optimization of machine learning algorithms, in *NeurIPS 25* (Curran Associates, Inc.), pp. 2951–2959.

Sohankar, J., Sadeghi, K., Banerjee, A., and Gupta, S. K. (2015). E-bias: A pervasive eeg-based identification and authentication system, in *Proceedings of the 11th ACM Symposium on QoS and Security for Wireless and Mobile Networks* (ACM), pp. 165–172.

Solon, A. J., Gordon, S. M., Lance, B., and Lawhern, V. (2017). Deep learning approaches for p300 classification in image triage: Applications to the nails task, in *Proceedings of the 13th NTCIR Conference on Evaluation of Information Access Technologies, NTCIR-13, Tokyo, Japan*, pp. 5–8.

Sønderby, C. K., Raiko, T., Maaløe, L., Sønderby, S. K., and Winther, O. (2016). Ladder variational autoencoders, in *Advances in Neural Information Processing Systems (NIPS)*.

Song, T., Zheng, W., Song, P., and Cui, Z. (2018). Eeg emotion recognition using dynamical graph convolutional neural networks, *IEEE Transactions on Affective Computing*.

Sors, A., Bonnet, S., Mirek, S., Vercueil, L., and Payen, J.-F. (2018). A convolutional neural network for sleep stage scoring from raw single-channel eeg, *Biomedical Signal Processing and Control* **42**, pp. 107–114.

Spampinato, C., Palazzo, S., Kavasidis, I., Giordano, D., Souly, N., and Shah, M. (2017). Deep learning human mind for automated visual classification, in *Proceedings of the IEEE Conference on Computer Vision and Pattern Recognition*, pp. 6809–6817.

Speier, W., Chandravadia, N., Roberts, D., Pendekanti, S., and Pouratian, N. (2017). Online bci typing using language model classifiers by als patients in their homes, *Brain-Computer Interfaces* **4**, 1-2, pp. 114–121.

St-Yves, G. and Naselaris, T. (2018). Generative adversarial networks conditioned on brain activity reconstruct seen images, in *2018 IEEE International Conference on Systems, Man, and Cybernetics (SMC)* (IEEE), pp. 1054–1061.

Stefano Filho, C. A., Attux, R., and Castellano, G. (2017). Eeg sensorimotor rhythms' variation and functional connectivity measures during motor imagery: Linear relations and classification approaches, *PeerJ* **5**, p. e3983.

Sternin, A., Stober, S., Grahn, J., and Owen, A. (2015). Tempo estimation from the eeg signal during perception and imagination of music, in *1st International Workshop on Brain-Computer Music Interfacing/11th International Symposium on Computer Music Multidisciplinary Research (BCMI/CMMR'15)(Plymouth)*.

Stober, S., Cameron, D. J., and Grahn, J. A. (2014a). Classifying eeg recordings of rhythm perception. in *ISMIR*, pp. 649–654.

Stober, S., Cameron, D. J., and Grahn, J. A. (2014b). Using convolutional neural networks to recognize rhythm stimuli from electroencephalography recordings, in *Advances in Neural Information Processing Systems*, pp. 1449–1457.

Stober, S., Sternin, A., Owen, A. M., and Grahn, J. A. (2015). Deep feature learning for eeg recordings, *arXiv preprint arXiv:1511.04306*.

Sturm, I., Lapuschkin, S., Samek, W., and Müller, K.-R. (2016). Interpretable deep neural networks for single-trial eeg classification, *Journal of Neuroscience Methods* **274**, pp. 141–145.

Suhaimi, N. F. M., Htike, Z. Z., and Rashid, N. K. A. M. (2015). Studies on classification of fmri data using deep learning approach.

Suk, H.-I., Shen, D., Initiative, A. D. N. *et al.* (2015). Deep learning in diagnosis of brain disorders, in *Recent Progress in Brain and Cognitive Engineering* (Springer), pp. 203–213.

Suk, H.-I., Wee, C.-Y., Lee, S.-W., and Shen, D. (2016). State-space model with deep learning for functional dynamics estimation in resting-state fmri, *NeuroImage* **129**, pp. 292–307.

Sun, B., Wang, Y., and Banda, J. (2014). Gait characteristic analysis and identification based on the iphone's accelerometer and gyrometer, *Sensors* **14**, 9, pp. 17037–17054.

Sun, S. (2008). The extreme energy ratio criterion for eeg feature extraction, in *International Conference on Artificial Neural Networks* (Springer), pp. 919–928.

Sundermeyer, M., Schlüter, R., and Ney, H. (2012). Lstm neural networks for language modeling, in *Thirteenth Annual Conference of the International Speech Communication Association*.

Supratak, A., Dong, H., Wu, C., and Guo, Y. (2017). Deepsleepnet: A model for automatic sleep stage scoring based on raw single-channel eeg, *IEEE Transactions on Neural Systems and Rehabilitation Engineering* **25**, 11, pp. 1998–2008.

Szczuko, P. (2017). Real and imaginary motion classification based on rough set analysis of eeg signals for multimedia applications, *Multimedia Tools and Applications* **76**, 24, pp. 25697–25711.

Tabar, Y. R. and Halici, U. (2016). A novel deep learning approach for classification of eeg motor imagery signals, *Journal of Neural Engineering* **14**, 1, p. 016003.

Taguchi, G. (1987). *System of Experimental Design: Engineering Methods to Optimize Quality and Minimize Costs* (UNIPUB/Kraus International Publications).

Talathi, S. S. (2017). Deep recurrent neural networks for seizure detection and early seizure detection systems, *arXiv preprint arXiv:1706.03283*.

Tan, C., Sun, F., and Zhang, W. (2018). Deep transfer learning for eeg-based brain computer interface, in *2018 IEEE International Conference on Acoustics, Speech and Signal Processing (ICASSP)* (IEEE), pp. 916–920.

Tan, C., Sun, F., Zhang, W., Chen, J., and Liu, C. (2017). Multimodal classification with deep convolutional-recurrent neural networks for electroencephalography, in *International Conference on Neural Information Processing* (Springer), pp. 767–776.

Tan, D., Zhao, R., Sun, J., and Qin, W. (2015a). Sleep spindle detection using deep learning: A validation study based on crowdsourcing, in *Engineering in Medicine and Biology Society (EMBC), 2015 37th Annual International Conference of the IEEE* (IEEE), pp. 2828–2831.

Tan, M., Santos, C. d., Xiang, B., and Zhou, B. (2015b). Lstm-based deep learning models for non-factoid answer selection, *arXiv preprint arXiv:1511.04108*.

Tang, Z., Li, C., and Sun, S. (2017). Single-trial eeg classification of motor imagery using deep convolutional neural networks, *Optik-International Journal for Light and Electron Optics* **130**, pp. 11–18.

Temko, A., Lightbody, G., Thomas, E. M., Boylan, G. B., and Marnane, W. (2011). Instantaneous measure of eeg channel importance for improved patient-adaptive neonatal seizure detection, *IEEE Transactions on Biomedical Engineering* **59**, 3, pp. 717–727.

Teo, J., Hou, C. L., and Mountstephens, J. (2017). Deep learning for eeg-based preference classification, in *AIP Conference Proceedings*, Vol. 1891 (AIP Publishing), p. 020141.

Thodoroff, P., Pineau, J., and Lim, A. (2016). Learning robust features using deep learning for automatic seizure detection, in *Machine Learning for Healthcare Conference*, pp. 178–190.

Thomas, J., Maszczyk, T., Sinha, N., Kluge, T., and Dauwels, J. (2017). Deep learning-based classification for brain-computer interfaces, in *Systems, Man, and Cybernetics (SMC), 2017 IEEE International Conference on* (IEEE), pp. 234–239.

Thomas, K. P. and Vinod, A. (2017). Eeg-based biometric authentication using gamma band power during rest state, *Circuits, Systems, and Signal Processing*, pp. 1–13.

Thomas, K. P. and Vinod, A. P. (2016a). Biometric identification of persons using sample entropy features of eeg during rest state, in *SMC* (IEEE), pp. 003487–003492.

Thomas, K. P. and Vinod, A. P. (2016b). Utilizing individual alpha frequency and delta band power in eeg based biometric recognition, in *Systems, Man, and Cybernetics (SMC), 2016 IEEE International Conference on* (IEEE), pp. 004787–004791.

Tokic, M. (2010). Adaptive ε-greedy exploration in reinforcement learning based on value differences, in *Annual Conference on Artificial Intelligence* (Springer), pp. 203–210.

Trejo, L. J., Kubitz, K., Rosipal, R., Kochavi, R. L., and Montgomery, L. D. (2015). Eeg-based estimation and classification of mental fatigue, *Psychology* **6**, 05, p. 572.

Tsinalis, O., Matthews, P. M., Guo, Y., and Zafeiriou, S. (2016). Automatic sleep stage scoring with single-channel eeg using convolutional neural networks, *arXiv preprint arXiv:1610.01683*.

Tsiouris, K. M., Pezoulas, V. C., Zervakis, M., Konitsiotis, S., Koutsouris, D. D., and Fotiadis, D. I. (2018). A long short-term memory deep learning network for the prediction of epileptic seizures using eeg signals, *Computers in Biology and Medicine* **99**, pp. 24–37.

Tu, T., Koss, J., and Sajda, P. (2018). Relating deep neural network representations to eeg-fmri spatiotemporal dynamics in a perceptual decision-making task, in *Proceedings of the IEEE Conference on Computer Vision and Pattern Recognition Workshops*, pp. 1985–1991.

Tu, W. and Sun, S. (2012). A subject transfer framework for eeg classification, *Neurocomputing* **82**, pp. 109–116.

Turner, J., Page, A., Mohsenin, T., and Oates, T. (2014). Deep belief networks used on high resolution multichannel electroencephalography data for seizure detection, in *2014 AAAI Spring Symposium Series*.

Ukil, A. (2006). Practical denoising of meg data using wavelet transform, in *International Conference on Neural Information Processing* (Springer), pp. 578–585.

Uktveris, T. and Jusas, V. (2017). Application of convolutional neural networks to four-class motor imagery classification problem, *Information Technology and Control* **46**, 2, pp. 260–273.

Ullah, I., Hussain, M., Aboalsamh, H. *et al.* (2018). An automated system for epilepsy detection using eeg brain signals based on deep learning approach, *Expert Systems with Applications* **107**, pp. 61–71.

Unar, J., Seng, W. C., and Abbasi, A. (2014). A review of biometric technology along with trends and prospects, *Pattern Recognition* **47**, 8, pp. 2673–2688.

Vallabhaneni, A., Wang, T., and He, B. (2005). Brain computer interface, in *Neural Engineering* (Springer), pp. 85–121.

Van Engelen, J. E. and Hoos, H. H. (2020). A survey on semi-supervised learning, *Machine Learning* **109**, 2, pp. 373–440.

Vařeka, L. and Mautner, P. (2017). Stacked autoencoders for the p300 component detection, *Frontiers in Neuroscience* **11**, p. 302.

Vaswani, A., Shazeer, N., Parmar, N., Uszkoreit, J., Jones, L., Gomez, A. N., Kaiser, L., and Polosukhin, I. (2017). Attention is all you need, in *Advances in Neural Information Processing Systems*, pp. 5998–6008.

Veeriah, V., Durvasula, R., and Qi, G.-J. (2015). Deep learning architecture with dynamically programmed layers for brain connectome prediction, in *SIGKDD* (ACM), pp. 1205–1214.

Vieira, S., Pinaya, W. H., and Mechelli, A. (2017). Using deep learning to investigate the neuroimaging correlates of psychiatric and neurological disorders: Methods and applications, *Neuroscience & Biobehavioral Reviews* **74**, pp. 58–75.

Vilamala, A., Madsen, K. H., and Hansen, L. K. (2017). Neural networks for interpretable analysis of eeg sleep stage scoring, in *International Workshop on Machine Learning for Signal Processing 2017.*

Völker, M., Schirrmeister, R. T., Fiederer, L. D., Burgard, W., and Ball, T. (2018). Deep transfer learning for error decoding from non-invasive eeg, in *Brain-Computer Interface (BCI), 2018 6th International Conference on* (IEEE), pp. 1–6.

Walker, J., Doersch, C., Gupta, A., and Hebert, M. (2016). An uncertain future: Forecasting from static images using variational autoencoders, in *European Conference on Computer Vision* (Springer).

Wang, B., Liu, K., and Zhao, J. (2016a). Inner attention based recurrent neural networks for answer selection, in *Proceedings of the 54th Annual Meeting of the Association for Computational Linguistics (Volume 1: Long Papers)*, Vol. 1, pp. 1288–1297.

Wang, F., Zhong, S.-h., Peng, J., Jiang, J., and Liu, Y. (2018). Data augmentation for eeg-based emotion recognition with deep convolutional neural networks, in *International Conference on Multimedia Modeling* (Springer), pp. 82–93.

Wang, H., Zhang, C., Shi, T., Wang, F., and Ma, S. (2015a). Real-time eeg-based detection of fatigue driving danger for accident prediction, *International Journal of Neural Systems* **25**, 02, p. 1550002.

Wang, K., Zhao, Y., Xiong, Q., Fan, M., Sun, G., Ma, L., and Liu, T. (2016b). Research on healthy anomaly detection model based on deep learning from multiple time-series physiological signals, *Scientific Programming* **2016**.

Wang, Q., Hu, Y., and Chen, H. (2017). Multi-channel eeg classification based on fast convolutional feature extraction, in *International Symposium on Neural Networks* (Springer), pp. 533–540.

Wang, X., Gong, G., Li, N., and Ma, Y. (2016c). A survey of the bci and its application prospect, in *Theory, Methodology, Tools and Applications for Modeling and Simulation of Complex Systems* (Springer), pp. 102–111.

Wang, Y., Gao, S., and Gao, X. (2006). Common spatial pattern method for channel selelction in motor imagery based brain-computer interface, in *Engineering in Medicine and Biology Society, 2005. IEEE-EMBS 2005. 27th Annual International Conference of the* (IEEE), pp. 5392–5395.

Wang, Z., Schaul, T., Hessel, M., Van Hasselt, H., Lanctot, M., and De Freitas, N. (2015b). Dueling network architectures for deep reinforcement learning, *arXiv preprint arXiv:1511.06581.*

Waytowich, N. R., Lawhern, V., Garcia, J. O., Cummings, J., Faller, J., Sajda, P., and Vettel, J. M. (2018). Compact convolutional neural networks for classification of asynchronous steady-state visual evoked potentials, *arXiv preprint arXiv:1803.04566.*

Wen, D., Wei, Z., Zhou, Y., Li, G., Zhang, X., and Han, W. (2018). Deep learning methods to process fmri data and their application in the diagnosis of cognitive impairment: A brief overview and our opinion, *Frontiers in Neuroinformatics* **12**, p. 23.

Wen, T. and Zhang, Z. (2018). Deep convolution neural network and autoencoders-based unsupervised feature learning of eeg signals, *IEEE Access* **6**, pp. 25399–25410.

West, J., Ventura, D., and Warnick, S. (2007). Spring research presentation: A theoretical foundation for inductive transfer, *Brigham Young University, College of Physical and Mathematical Sciences* **1**, 08.

Wilaiprasitporn, T., Ditthapron, A., Matchaparn, K., Tongbuasirilai, T., Banluesombatkul, N., and Chuangsuwanich, E. (2019). Affective eeg-based person identification using the deep learning approach, *IEEE Transactions on Cognitive and Developmental Systems*.

Wu, Z., Pan, S., Chen, F., Long, G., Zhang, C., and Philip, S. Y. (2020). A comprehensive survey on graph neural networks, *IEEE Transactions on Neural Networks and Learning Systems*.

Wulsin, D., Blanco, J., Mani, R., and Litt, B. (2010). Semi-supervised anomaly detection for eeg waveforms using deep belief nets, in *2010 Ninth International Conference on Machine Learning and Applications* (IEEE), pp. 436–441.

Xia, B., Li, Q., Jia, J., Wang, J., Chaudhary, U., Ramos-Murguialday, A., and Birbaumer, N. (2015). Electrooculogram based sleep stage classification using deep belief network, in *Neural Networks (IJCNN), 2015 International Joint Conference on* (IEEE), pp. 1–5.

Xie, Z., Schwartz, O., and Prasad, A. (2018). Decoding of finger trajectory from ecog using deep learning, *Journal of Neural Engineering* **15**, 3, p. 036009.

Xu, H. and Plataniotis, K. N. (2016a). Affective states classification using eeg and semi-supervised deep learning approaches, in *Multimedia Signal Processing (MMSP), 2016 IEEE 18th International Workshop on* (IEEE), pp. 1–6.

Xu, H. and Plataniotis, K. N. (2016b). Eeg-based affect states classification using deep belief networks, in *Digital Media Industry & Academic Forum (DMIAF)* (IEEE), pp. 148–153.

Yang, C., Lieberman, J., and Hong, C. (1989). Early smooth horizontal eye movement: A favorable prognostic sign in patients with locked-in syndrome, *Archives of Physical Medicine and Rehabilitation* **70**, 3, pp. 230–232.

Yang, H., Sakhavi, S., Ang, K. K., and Guan, C. (2015). On the use of convolutional neural networks and augmented csp features for multi-class motor imagery of eeg signals classification, in *Engineering in Medicine and Biology Society (EMBC), 2015 37th Annual International Conference of the IEEE* (IEEE), pp. 2620–2623.

Yang, S. and Deravi, F. (2014). Novel hht-based features for biometric identification using eeg signals, in *ICPR* (IEEE), pp. 1922–1927.

Yano, V., Zimmer, A., and Ling, L. L. (2012). Multimodal biometric authentication based on iris pattern and pupil light reflex, in *Pattern Recognition (ICPR), 2012 21st International Conference on* (IEEE), pp. 2857–2860.

Yao, L., Nie, F., Sheng, Q. Z., Gu, T., Li, X., and Wang, S. (2016). Learning from less for better: Semi-supervised activity recognition via shared structure discovery, in *UbiComp* (ACM).

Yao, L., Sheng, Q. Z., Li, X., Gu, T., Tan, M., Wang, X., Wang, S., and Ruan, W. (2018). Compressive representation for device-free activity recognition with passive rfid signal strength, *IEEE Transactions on Mobile Computing* **17**, 2, pp. 293–306.

Yepes, A. J., Tang, J., and Mashford, B. S. (2017). Improving classification accuracy of feedforward neural networks for spiking neuromorphic chips, *arXiv preprint arXiv:1705.07755*.

Yin, E., Zeyl, T., Saab, R., Chau, T., Hu, D., and Zhou, Z. (2015). A hybrid brain–computer interface based on the fusion of p300 and ssvep scores, *IEEE Transactions on Neural Systems and Rehabilitation Engineering* **23**, 4, pp. 693–701.

Yin, Z. and Zhang, J. (2017). Cross-session classification of mental workload levels using eeg and an adaptive deep learning model, *Biomedical Signal Processing and Control* **33**, pp. 30–47.

Yin, Z., Zhao, M., Wang, Y., Yang, J., and Zhang, J. (2017). Recognition of emotions using multimodal physiological signals and an ensemble deep learning model, *Computer Methods and Programs in Biomedicine* **140**, pp. 93–110.

Yoon, J., Lee, J., and Whang, M. (2018). Spatial and time domain feature of erp speller system extracted via convolutional neural network, *Computational Intelligence and Neuroscience* **2018**.

Yoshida, K., Li, X., Cano, G., Lazarus, M., and Saper, C. B. (2009). Parallel preoptic pathways for thermoregulation, *Journal of Neuroscience* **29**, 38, pp. 11954–11964.

Yuan, Y., Xun, G., Jia, K., and Zhang, A. (2017). A novel wavelet-based model for eeg epileptic seizure detection using multi-context learning, in *Bioinformatics and Biomedicine (BIBM), 2017 IEEE International Conference on* (IEEE), pp. 694–699.

Yuan, Y., Xun, G., Ma, F., Suo, Q., Xue, H., Jia, K., and Zhang, A. (2018). A novel channel-aware attention framework for multi-channel eeg seizure detection via multi-view deep learning, in *Biomedical & Health Informatics (BHI), 2018 IEEE EMBS International Conference on* (IEEE), pp. 206–209.

Zaremba, W., Sutskever, I., and Vinyals, O. (2014). Recurrent neural network regularization, *arXiv preprint arXiv:1409.2329*.

Zeng, H., Wu, Z., Zhang, J., Yang, C., Zhang, H., Dai, G., and Kong, W. (2019). Eeg emotion classification using an improved sincnet-based deep learning model, *Brain Sciences* **9**, 11, p. 326.

Zhang, D., Yao, L., Zhang, X., Wang, S., Chen, W., and Boots, R. (2018a). Eeg-based intention recognition from spatio-temporal representations via cascade and parallel convolutional recurrent neural networks, *The 32nd AAAI Conference on Artificial Intelligence (AAAI 2018)*.

Zhang, J., Wu, Y., Bai, J., and Chen, F. (2016a). Automatic sleep stage classification based on sparse deep belief net and combination of multiple classifiers, *Transactions of the Institute of Measurement and Control* **38**, 4, pp. 435–451.

Zhang, K., Zuo, W., Gu, S., and Zhang, L. (2017a). Learning deep cnn denoiser prior for image restoration, in *CVPR*, pp. 3929–3938.

Zhang, T., Zheng, W., Cui, Z., Zong, Y., and Li, Y. (2018b). Spatial-temporal recurrent neural network for emotion recognition, *IEEE Transactions on Cybernetics*, 99, pp. 1–9.

Zhang, X., Chen, X., Yao, L., Ge, C., and Dong, M. (2019a). Deep neural network hyperparameter optimization with orthogonal array tuning, *The 26th International Conference On Neural Information Processing (ICONIP 2019)*.

Zhang, X. and Wu, D. (2019). On the vulnerability of cnn classifiers in eeg-based bcis, *IEEE Transactions on Neural Systems and Rehabilitation Engineering* **27**, 5, pp. 814–825.

Zhang, X., Yao, L., Huang, C., Gu, T., Yang, Z., and Liu, Y. (2019b). Deepkey: A multimodal biometric authentication system via deep decoding gaits and brainwaves, *ACM Transaction on Intelligent Systems and Technology (TIST)*.

Zhang, X., Yao, L., Huang, C., Kanhere, S. S., and Zhang, D. (2018c). Brain2object: Printing your mind from brain signals with spatial correlation embedding, *Pattern Recognition*.

Zhang, X., Yao, L., Huang, C., Sheng, Q. Z., and Wang, X. (2017b). Intent recognition in smart living through deep recurrent neural networks, in *International Conference on Neural Information Processing (ICONIP 2017)* (Springer), pp. 748–758.

Zhang, X., Yao, L., Huang, C., Wang, S., Tan, M., Long, G., and Wang, C. (2018d). Multi-modality sensor data classification with selective attention, in *Proceedings of the 27th International Joint Conference on Artificial Intelligence (IJCAI 2018)*.

Zhang, X., Yao, L., Huang, C., Wang, S., Tan, M., Long, G., and Wang, C. (2018e). Multi-modality sensor data classification with selective attention, in *IJCAI-18*, pp. 3111–3117.

Zhang, X., Yao, L., Huang, C., Wang, S., Tan, M., Long, G., and Wang, C. (2018f). Multi-modality sensor data classification with selective attention, in *The 27th International Joint Conference on Artificial Intelligence, IJCAI-18*, pp. 3111–3117.

Zhang, X., Yao, L., Kanhere, S. S., Liu, Y., Gu, T., and Chen, K. (2018g). MindID: Person identification from brain waves through attention-based recurrent neural network, *the ACM International Conference on Pervasive and Ubiquitous Computing (UbiComp 2018)* **2**, 3, p. 149.

Zhang, X., Yao, L., Sheng, Q. Z., Kanhere, S. S., Gu, T., and Zhang, D. (2018h). Converting your thoughts to texts: Enabling brain typing via deep feature learning of eeg signals, in *2018 IEEE International Conference on Pervasive Computing and Communications (PerCom 2018)* (IEEE), pp. 1–10.

Zhang, X., Yao, L., Wang, X., Monaghan, J., Mcalpine, D., and Zhang, Y. (2019c). A survey on deep learning based brain computer interface: Recent advances and new frontiers, *arXiv preprint arXiv:1905.04149*.

Zhang, X., Yao, L., Wang, X., Zhang, W., Zhang, S., and Liu, Y. (2019d). Know your mind: Adaptive brain signal classification with reinforced attentive convolutional neural networks, in *The 2019 IEEE International Conference on Data Mining (ICDM 2019)*.

Zhang, X., Yao, L., and Yuan, F. (2019e). Adversarial variational embedding for robust semi-supervised learning, *The 25th ACM SIGKDD Conference on Knowledge Discovery and Data Mining (KDD 2019)*.

Zhang, X., Yao, L., Zhang, D., Wang, X., Sheng, Q. Z., and Gu, T. (2017c). Multi-person brain activity recognition via comprehensive eeg signal analysis, in *Proceedings of the 14th EAI International Conference on Mobile and Ubiquitous Systems: Computing, Networking and Services (MobiQuitous 2017)* (ACM), pp. 28–37.

Zhang, X., Yao, L., Zhang, S., Kanhere, S., Sheng, M., and Liu, Y. (2018i). Internet of things meets brain–computer interface: A unified deep learning framework for enabling human-thing cognitive interactivity, *IEEE Internet of Things Journal* **6**, 2, pp. 2084–2092.

Zhang, Y., Roller, S., and Wallace, B. (2016b). Mgnc-cnn: A simple approach to exploiting multiple word embeddings for sentence classification, *arXiv preprint arXiv:1603.00968*.

Zhang, Y., Yang, S., Liu, Y., Zhang, Y., Han, B., and Zhou, F. (2018j). Integration of 24 feature types to accurately detect and predict seizures using scalp eeg signals, *Sensors (Basel, Switzerland)* **18**, 5.

Zhao, Y. and He, L. (2014). Deep learning in the eeg diagnosis of alzheimer's disease, in *Asian Conference on Computer Vision* (Springer), pp. 340–353.

Zheng, W.-L., Guo, H.-T., and Lu, B.-L. (2015). Revealing critical channels and frequency bands for emotion recognition from eeg with deep belief network, in *Neural Engineering (NER), 2015 7th International IEEE/EMBS Conference on* (IEEE), pp. 154–157.

Zheng, W.-L. and Lu, B.-L. (2015). Investigating critical frequency bands and channels for eeg-based emotion recognition with deep neural networks, *IEEE Transactions on Autonomous Mental Development* **7**, 3, pp. 162–175.

Zheng, W.-L. and Lu, B.-L. (2016). Personalizing eeg-based affective models with transfer learning, in *Proceedings of the Twenty-Fifth International Joint Conference on Artificial Intelligence* (AAAI Press), pp. 2732–2738.

Zheng, W.-L., Zhu, J.-Y., Peng, Y., and Lu, B.-L. (2014). Eeg-based emotion classification using deep belief networks, in *Multimedia and Expo (ICME), 2014 IEEE International Conference on* (IEEE), pp. 1–6.

Zhou, M., Tian, C., Cao, R., Wang, B., Niu, Y., Hu, T., Guo, H., and Xiang, J. (2018). Epileptic seizure detection based on eeg signals and cnn, *Frontiers in Neuroinformatics* **12**, p. 95.

Zhou, P., Shi, W., Tian, J., Qi, Z., Li, B., Hao, H., and Xu, B. (2016). Attention-based bidirectional long short-term memory networks for relation classification, in *ACL (Volume 2: Short Papers)*, pp. 207–212.

Zitnik, M., Nguyen, F., Wang, B., Leskovec, J., Goldenberg, A., and Hoffman, M. M. (2019). Machine learning for integrating data in biology and medicine: Principles, practice, and opportunities, *Information Fusion* **50**, pp. 71–91.

Ziyabari, S., Shah, V., Golmohammadi, M., Obeid, I., and Picone, J. (2017). Objective evaluation metrics for automatic classification of eeg events, *arXiv preprint arXiv:1712.10107*.

Index

Note: Page numbers followed by f, or t indicate material in figures, or tables respectively.

A

AASM. *See* American Academy of Sleep Medicine
Accuracy, 142
ACGAN, 206, 208–209
AD. *See* Alzheimer's disease
Adagrad. *See* Adaptive sub-gradient method
Adam. *See* Adaptive moment estimation
Adam optimizer, 157
Adaptive moment estimation (Adam), 31
Adaptive sub-gradient method (Adagrad), 31
Adversarial variational embedding (AVAE) framework, 151, 154f, 156–157, 161, 239
AE. *See* Autoencoder
AEP. *See* Auditory evoked potential
Affective computing, deep learning-based BCI systems in, 86, 92
Agent policy/optimization, 131–133
Algorithm, 126, 134, 139, 148, 158, 181, 188

Alpha pattern (8–12Hz), 16
Alzheimer's disease (AD), 62–63
    detection of, 91–92
American Academy of Sleep Medicine (AASM), 48
American Electroencephalographic Society, 14
Architecture, 206–207
Area Under the Curve (AUC), 143
Attention mechanism, 124, 237–238
AUC. *See* Area Under the Curve
Auditory evoked potential (AEP), 23, 58–59
Authentication
    EEG-based person identification, 169–182
        challenges, 169–173, 171t
        EEG pattern analysis, 173–175, 174f, 175t
        methodology, 175–181, 176f
    person authentication, 182–190
        data acquisition, 189–190, 190f
        methodology, 184–189, 186–187f
        motivations, 183–184, 185t

Autoencoder (AE), 28, 36–38, 36f
AVAE framework. *See* Adversarial
    variational embedding framework

B

BCI system. *See* Brain-computer
    interface system
Benchmark data sets, 88–91, 89–90t
Beta pattern (12–30Hz), 16
Blood-oxygen-level-dependent
    (BOLD) response, 17–18
BOLD response. *See*
    Blood-oxygen-level-dependent
    response
Brain-computer interface (BCI)
    system
        background on, 3–5
        categories of, 97
        general workflow of, 3–4, 4f
Brain-controlled typing system,
    216–219, 217–218f, 217t
Brain signal
        decomposition, 98
        discussions on, 71–73
        publication proportions for, 71f
        representations, taxonomy of,
            98–99
        terminology, 99–100
Brain signal acquisition methods,
    9–26, 10f, 11t
        EEG paradigms, 21–26
            evoked potentials, 21–26
            spontaneous, 21, 22
        invasive approach, 9–10, 12–13,
            12f
            electrocorticography, 13, 13f
            intracortical recording
                technique, 12, 12f
        noninvasive approach, 14–21
            electroencephalography,
                14–17, 15f, 16t
            electrooculography, 14, 19–20,
                20f
            functional magnetic resonance
                imaging, 14, 18, 19f

functional near-infrared
    spectroscopy, 14, 17–18,
    17f
magnetoencephalography, 14,
    20–21, 21f

C

CA hybrid models. *See*
    Classification-aimed hybrid models
Cascade R-CNN model, 110, 111f
CGAN, 203, 208
Classification-aimed (CA) hybrid
    models, 43–44
Classification component, in BCI
    workflow, 5
CNN. *See* Convolutional neural
    networks
Collection method, in BCI workflow,
    4
Common Spatial Pattern (CSP),
    191–195
Communication, deep learning-based
    BCI systems in, 85
Concealed information test, deep
    learning-based BCI systems in, 88
Convolutional neural networks
    (CNN), 28, 32f, 34–35, 98–99,
    103–105, 103f, 109–110, 121,
    162–163, 192, 196–197, 204
Cross-scenario classification
        attention-based classification
            across applications, 135–147
            evaluation across applications,
                138–147, 144t, 145f, 146t
            reinforced attentive CNN,
                136–138, 137f
        attention-based classification
            across signal sources, 124–135
            data replication and shuffling,
                128
            motivation, 126
            reinforced selective attention
                model, 125–126, 127f
            selective attention mechanism,
                128–133, 130f

transfer learning methods, 147–148
Cross-similarity, defined, 114
Cross-subject BCI system, 97
   EEG characteristic analysis, 113–116, 114*t*
   feature representation, 118–119
   intent recognition, 119–120
   overview, 113
   representation learning framework, 115, 117–120, 117*f*
   normalization, 115, 117–118
Cross-subject correlation coefficient matrix, 115, 116*t*
CSP. *See* Common Spatial Pattern

D

Data acquisition experiment, 198*f*
Data augmentation, 55
Data transformation procedure, 216
DBN. *See* Deep belief networks
DCGAN. *See* Deep convolutional GAN
Deep belief networks (DBN), 28, 39–40, 39*f*, 165
Deep convolutional GAN (DCGAN), 63
Deep feature learning, language interpretation, 212–215
   CNN feature learning, 214–215
   RNN feature learning, 212–214, 213*f*
Deep learning-based BCI systems, 47–75
   applications, 77–93, 78–83*t*
      affective computing, 86, 92
      benchmark data sets, 88–91, 89–90*t*
      communication, 85
      discussions, 91–93
      driver fatigue detection, 86–87, 93
      health care, 77, 84
      mental load measurement, 87

security, 85–86, 92
   smart environment, 84
   discussion, 64–75, 65–70*t*, 71*f*
      on brain signals, 71–73
      on deep learning models, 73–75
   ECoG brain signals, 47–48
   EEG potentials, 48–61
      evoked potentials, 58–61
      spontaneous, 48–57
   EOG, 63–64
   fMRI, 62–63
   fNIRS, 61
   future directions, 237–240
      general framework, 237–238
      hardware portability, 239
      semi-supervised classification, 238–239
      subject-independent classification, 238
      unsupervised classification, 238–239
   intracortical brain signals, 47–48
   MEG, 64
   summary of, 65–70*t*
Deep learning models, 5–6, 27–44, 28*f*
   discriminative, 27–35
      convolutional neural networks, 28, 32*f*, 34–35
      multilayer perceptron, 28–31
      recurrent neural networks, 28, 31–34, 32*f*
   discussions on, 73–75
   generative, 28, 40–42, 43*f*
      generative adversarial networks, 28, 42, 43*f*
      variational autoencoder, 28, 40–42, 43*f*
   hybrid, 28, 43–44
   publication proportions for, 71*f*
   representative, 28, 35–40
      autoencoder, 28, 36–38, 36*f*
      deep belief networks, 28, 39–40, 39*f*

restricted Boltzmann
machine, 28, 36*f*, 38–39
types, 27–28, 29*t*
Deep Q networks (DQNs), 131
DeepKey, 183–184, 186, 189
Delta pattern (0.5–4Hz), 16
DGR. *See* Dynamic graph
representation
Discriminative deep learning models,
27–35
convolutional neural networks,
32*f*, 34–35
multilayer perceptron, 29–31
recurrent neural networks,
31–34, 32*f*
emotional EEG signals, 52
fMRI in, 62
mental disease EEG signals,
53–54
MI EEG signals, 50
sleep EEG signals, 49
DQNs. *See* Deep Q networks
Driver fatigue detection, deep
learning-based BCI systems in,
86–87, 93
Dynamic graph representation
(DGR), 191, 196–197

E

ECoG. *See* Electrocorticography
ECoG brain signals, 47–48
EEG. *See* Electroencephalography
EEG characteristic analysis, 113–116,
114*t*
in cross-subject situation, 114
in interclass situation, 114
EEG motor movement/imagery
database (EEGMMIDB), 140–141
EEG potentials, 48–61
evoked potentials, 58–61
ERP, 58–60
SSEP, 60–61
spontaneous, 48–57
data augmentation, 55

emotional, 51–53
mental disease, 53–54
motor-imagery, 50–51
other models, 55–57
sleep, 48–49
EEG-S, 141
EEGMMIDB. *See* EEG motor
movement/imagery database
EEGNET, 105
EER. *See* Extreme Energy Ratio
Electrocorticography (ECoG), 13,
13*f*
Electroencephalography (EEG),
14–17, 15*f*, 16*t*
paradigms, 21–26
evoked potentials, 21–26
spontaneous, 21–22
Electrooculography (EOG), 14,
19–20, 20*f*, 63–64
Emotional EEG, 51–53
Emotiv headset, 188–189, 200–201
Encoder-Decoder RNN, 177–178
EOG. *See* Electrooculography
Epilepsy, 227
EPs. *See* Evoked potentials
ERPs. *See* Event-related potentials
Event-related
desynchronization/synchronization
(ERD/ERS), 51
Event-related potentials (ERPs), 21,
22–23, 58–60
auditory evoked potentials, 23
somatosensory evoked
potentials, 23
visual evoked potentials, 23
AEP, 58–59
RSVP, 59–60
VEP, 58
Evoked potentials (EPs), 21–26,
58–61
event-related potential, 22–23
P300, 24–25
steady state evoked potentials,
25
visual-related potentials, 25–26
ERP, 58–60

AEP, 58–59
RSVP, 59–60
VEP, 58
SSEP, 60–61
Extreme Energy Ratio (EER), 122
Extreme Gradient Boosting. *See*
XGBoost model

F

F1 Score, 142
Feature engineering, defined, 5
Feature representation, 118–119
fMRI. *See* Functional magnetic
resonance imaging
fNIRS. *See* Functional near-infrared
spectroscopy
Functional magnetic resonance
imaging (fMRI), 14, 18, 19f, 62–63,
109, 109f, 203
Functional near-infrared spectroscopy
(fNIRS), 14, 17–18, 17f, 61

G

Gamma pattern (30–100Hz), 17
GANs. *See* Generative adversarial
networks
Gated recurrent unit (GRU), 33f, 34
GCN. *See* Graph Convolutional
Network
Gender detection, deep
learning-based BCI systems in, 88
Generative adversarial networks
(GANs), 28, 42, 43f, 151–152,
154f, 156, 161, 203–204, 206,
208–209
Generative deep learning models, 28,
40–42, 43f
generative adversarial networks,
28, 42, 43f
variational autoencoder, 28,
40–42, 43f
fMRI in, 63
GNNs. *See* Graph neural networks

Graph Convolutional Network
(GCN), 106–107, 108f
Graph neural networks (GNNs),
98–99, 106
GRU. *See* Gated recurrent unit
Guilty knowledge test, deep
learning-based BCI systems in, 88

H

Health care, deep learning-based BCI
systems in, 77, 84
Hybrid deep learning models, 28,
43–44
emotional EEG signals, 53
mental disease EEG signals, 54
MI EEG signals, 50–51
sleep EEG signals, 49

I

Intent recognition, 119–120,
221–226
deployment, 225–226, 225t
mind-controlled appliances,
226
mind-controlled mobile robot,
225, 226f
orthogonal array tuning method
overview, 222–223
workflow, 223–224, 224f
system overview, 221–222, 222f
Interclass correlation coefficient
matrix, 114, 114t
Intersubject transfer learning,
121–122
Intersubject variability, 97
Intracortical brain signals, 47–48
Invasive approach, in brain signal
acquisition, 9–10, 12–13, 12f
electrocorticography, 13, 13f
intracortical recording
technique, 12, 12f

K

Kullback–Leibler divergence, 155

L

Language interpretation, 211–219
  brain-controlled typing system, 216–219, 217–218*f*, 217*t*
  discussion, 219
  methodology, 212–216
    deep feature learning, 212–215
    feature adaptation, 215–216
    overview, 212
Long short-term memory (LSTM), 32–33, 33*f*, 101–102, 110, 126, 133, 178–179
Loss function, 207
LSTM. *See* Long short-term memory

M

Magnetoencephalography (MEG), 14, 20–21, 21*f*, 64
MCI. *See* Mild cognitive impairment
MEG. *See* Magnetoencephalography
Mental disease EEG, 53–54
Mental load measurement, deep learning-based BCI systems in, 87
Mild cognitive impairment (MCI), 63
Min–max normalization, 115, 117
Mind-controlled appliances, 226
Mind-controlled mobile robot, 225, 226*f*
MindID, 181–182, 184
MIR. *See* Movement Intention Recognition
MLP. *See* Multilayer perceptron
Motor-imagery EEG, 50–51
Movement Intention Recognition (MIR), 138–139
Multilayer perceptron (MLP), 28–31
  vs. standard neural network, 29–30, 30*f*

N

NCA hybrid models. *See* Nonclassification-aimed hybrid models
ND. *See* Neurological Diagnosis
Neurological Diagnosis (ND), 140

Nonclassification-aimed (NCA) hybrid models, 44
Noninvasive approach, in brain signal acquisition, 14–21
  electroencephalography, 14–17, 15*f*, 16*t*
  electrooculography, 14, 19–20, 20*f*
  functional magnetic resonance imaging, 14, 18, 19*f*
  functional near-infrared spectroscopy, 14, 17–18, 17*f*
  magnetoencephalography, 14, 20–21, 21*f*
Noninvasive EEG signals, 228

O

OATM. *See* Orthogonal array tuning method
Occipital lobe, 124
Orthogonal array tuning method (OATM)
  overview, 222–223
  workflow, 223–224, 224*f*
Overlap, 99–100

P

P300 (P3), 24–25
Parallelized R-CNN model, 111, 112*f*
Patient-independent epileptic seizure diagnosis method, 229, 230*f*, 231
Patient-independent neurological disorder detection, 227–236
  discussions, 235–236
  introduction, 227–229
    challenges, 227–228
    contributions, 229
    motivation, 228–229
  methodology, 229–235
    attention-based seizure diagnosis, 233–234
    EEG decomposition, 231–233
    overview, 229, 230*f*, 231

patient detection, 234
training details, 235
Person Identification (PI), 139
PET. *See* Positron emission
tomography
PI. *See* Person Identification
Pinkie Pie, 194
Pooling layer, 214–215
Positron emission tomography
(PET), 62
Precision, 142
Proposed approach, 137*f*, 172–173

R

Rapid serial visual presentation
(RSVP), 23, 59–60
RBM. *See* Restricted Boltzmann
machine
Recall, 142
Receiver Operating Characteristic
(ROC), 142–143, 145*f*
Recurrent neural networks (RNN),
28, 31–34, 32*f*, 109–110, 172–173,
189
gated recurrent unit, 33*f*, 34
long short-term memory, 32–33,
33*f*
Reinforcement learning, 238
Representation learning framework,
115, 117–120, 117*f*
normalization, 115, 117–118
min–max, 115, 117
unity, 117–118
z-score scaling, 118
Representative deep learning models,
28, 35–40
autoencoder, 28, 36–38, 36*f*
deep belief networks, 28, 39–40,
39*f*
restricted Boltzmann machine,
28, 36*f*, 38–39
Representative models
emotional EEG signals, 52–53
fMRI in, 62–63
mental disease EEG signals, 54

MI EEG signals, 50
sleep EEG signals, 49
ResNet, 103
Restricted Boltzmann machine
(RBM), 28, 36*f*, 38–39
Reward model, 130–131
RNN. *See* Recurrent neural networks
Robot Operating System (ROS), 225
ROC. *See* Receiver Operating
Characteristic
ROS. *See* Robot Operating System
RSVP. *See* Rapid serial visual
presentation

S

Sample, defined, 99
Sampling operation, 100
Security, deep learning-based BCI
systems in, 85–86, 92
Segmentation
defined, 99
key parameters in, 99–100
Segments, 99
Self-similarity, defined, 114
Semi-supervised classification
generative methods, 149–161,
150*f*
adversarial variational
embedding algorithm,
152–157, 154*f*
evaluation, 157–160, 159*t*,
160*f*
unsupervised representations
learning, 164–165
wrapper methods, 161–164
boosting, 164
co-training, 163–164
self-training, 162–163
Semi-supervised GAN, 239
SEP. *See* Somatosensory evoked
potentials
SGD. *See* Stochastic gradient descent
Signal preprocessing, in BCI
workflow, 4
Signal-to-noise ratio (SNR), 9–10

SincNet, 103
Sleep EEG, 48–49
Smart environment, deep
   learning-based BCI systems in,
   84
SNR. *See* Signal-to-noise ratio
Softmax function, 197
Somatosensory evoked potentials
   (SEP), 23
Spike, 12
Spontaneous EEG, 21–22, 48–57
      data augmentation, 55
      emotional, 51–53
      mental disease, 53–54
      motor-imagery, 50–51
      other models, 55–57
      sleep, 48–49
SSAEP. *See* Steady-state auditory
   evoked potentials
SSEP. *See* Steady-state evoked
   potentials
SSSEP. *See* Steady-state
   somatosensory evoked potentials
SSVEP. *See* Steady-state visually
   evoked potentials
Standard neural network vs.
   multilayer perceptron, 29–30, 30*f*
Standardization. *See Z*-score scaling
Steady-state auditory evoked
   potentials (SSAEP), 25
Steady-state evoked potentials
   (SSEP), 21, 25, 60–61
Steady-state somatosensory evoked
   potentials (SSSEP), 25
Steady-state visually evoked
   potentials (SSVEP), 25, 60–61
Stochastic gradient descent (SGD), 31
Subject-dependent BCI system, 97,
   100–112
      discussion, 112
      graphical representation
         learning, 105–109, 106*f*,
         108–109*f*
      spatial representation learning,
         103–105, 103*f*

spatiotemporal representation
   learning, 109–111, 110*f*, 112*f*
temporal representation
   learning, 100–103, 101*f*
Subject-independent BCI system, 97,
   120–122
      intersubject transfer learning,
         121–122
      transfer learning, 121
Supervision rate, 158–160

T

Temporal representation learning,
   100–103, 101*f*
Theta pattern (4–8 Hz), 16
Time point, 100
Time window, 99–100
Transfer learning, 121
TUH, 141–142, 158, 160*f*

U

Unity normalization, 117–118

V

VAE. *See* Variational autoencoder
VAE++ model, 150–155, 150*f*, 154*f*,
   156, 239
Variational autoencoder (VAE), 28,
   40–42, 43*f*, 150–156, 150*f*
VEP. *See* Visual evoked potential
Visual evoked potential (VEP), 58,
   23, 25–26
Visual object recognition, deep
   learning-based BCI systems in, 88
Visual reconstruction
      Brain2Object, 191–202, 193*f*
         data acquisition, 198–199,
            198*f*
         online system, 199–201,
            199–201*f*
         geometrical shape
            reconstruction, 202, 203*f*
         EEG signal acquisition,
            203–204

evaluations, 208–210, 209*f*,
    210*t*
  methodology, 204–208, 205*f*
Visualization, 160

W

Wasserstein GAN (WGAN), 63
WGAN. *See* Wasserstein GAN

X

XGBoost model, 113, 115, 119–120,
  180, 216

Z

Z-score scaling, 118

www.ingramcontent.com/pod-product-compliance
Lightning Source LLC
Chambersburg PA
CBHW050546190326
41458CB00007B/1934